RELIABILITY FOR 1

Other Macmillan titles of related interest

Mechanical Reliability, 2nd edition, A. D. S. Carter
Reliability and Maintainability in Perspective, 2nd edition, D. J. Smith

Reliability for Engineers

An Introduction

Eur Ing Michael Beasley
BSc (Hons), CEng, MIEE, FSaRS

First published 1991 by
MACMILLAN PRESS LTD
Houndmills, Basingstoke, Hampshire RG21 6XS
and London
Companies and representatives
throughout the world

ISBN 0–333–54237–1 hardback
ISBN 0–333–54238–X paperback

A catalogue record for this book is available
from the British Library.

10 9 8 7 6 5 4 3 2
03 02 01 00 99 98 97 96 95

Printed in Hong Kong

To my wife Frances and our children Ian and Susan

Contents

Preface

This book is the outcome of lectures on reliability which I have given, firstly in-house in STC plc and later at Hatfield Polytechnic. It is intended to help both students who are new to the subject and practising engineers who are looking for help in reliability engineering.

Until my retirement I spent my working life in the telecommunications industry, and thus I had electrical engineers mostly in mind when writing this text. However, the theory and principles of reliability engineering are of wide applicability so that I hope this book will also be useful to workers in other engineering disciplines.

It is difficult, if not impossible, to learn reliability engineering in complete isolation. So my warm thanks go to all my former colleagues in STC plc who by discussions and instruction helped my understanding of the subject; in particular I should like to mention Dr G. G. Pullum, Mr G. H. P. Breuer and Mr I. Campbell.

I have also been greatly helped as a part-time lecturer at Hatfield Polytechnic by my colleagues Mr E. A. De Maria, Mr P. J. Bunn and Mr P. F. Jackson, to all of whom I owe a debt of gratitude.

There are already a number of good textbooks on reliability; my excuse for writing yet another is that I believe mine to be of practical help - to the student as well as to the qualified engineer.

I am grateful to Mrs D. J. Ross and Mrs S. M. Pilkington who typed early versions of the lecture notes from which this book has been developed; and also to my wife Frances who had the arduous task of typing most of the final manuscript.

Finally, it is a pleasure to express my thanks to Mr Malcolm Stewart and his colleagues of Macmillan Education Ltd for their help in bringing this book to press.

1 Introduction

1.1 The Study of Reliability

Reliability (which is defined more precisely later) is the study of how, why and when failures occur. This book does not cover the physics of failure of components since this is a substantial field in its own right. Although the statistics of component failure are described, the main purpose of this book is to introduce the reader to the methods of specification, design, analysis, manufacture and testing of reliable systems. The systems envisaged are chiefly in the telecommunications field although the principles introduced here can be used much more widely. From our viewpoint the study of reliability needs a combination of electrical engineering (for system analysis), statistics (for the analysis of failure data), and probability theory (for the prediction of failure patterns).

We shall be concerned chiefly with system failure caused by component failure (either catastrophic or by deterioration) although software reliability is also briefly considered. An example of a component's catastrophic failure is a capacitor which goes open- or short-circuit. Deterioration of a component is exemplified by a pair of contacts in a relay which gradually develop a high contact resistance. When the contact resistance becomes excessive then the relay is considered to have failed.

It is important that the distinction between relevant and non-relevant failures should be clearly understood. As an example, it would be possible for a badly-constructed amplifier to oscillate because of the proximity of the input and output connections. An examination of the amplifier would not reveal a faulty component yet nevertheless the performance would be unsatisfactory because of poor design. The amplifier may loosely be called 'unreliable' but its unsatisfactory performance is due to design inadequacy rather than unreliability. Similarly a TV receiver which might be dubbed 'unreliable' may be giving an unsatisfactory picture because the signal strength is too low. The problem then lies with the siting of the aerial rather than improving the reliability of the receiver In a similar fashion a user may misunderstand the controls on a piece of equipment and declare it to be 'unreliable' because he cannot make it work properly. It is the role of Human Factors to ensure that the controls on a piece of equipment are easily understood.

Because so many issues are involved in the design and manufacture of 'good' systems, the study of reliability should be regarded as a part of System Effectiveness, the study of which developed historically from that of reliability.

1.2 Historical Survey

Ever since man first made objects, he has been faced with the problem of unreliability; shoes wore out, spears broke, bridges collapsed. Improvements were made partly by trial-and-error and partly by a simple process of analysis and re-design: if a part was of insufficient strength then it was made stronger. Even in ancient times this process led to some spectacular successes in the reliability field. If the builder of the Great Pyramid at Giza had been asked 'How reliable is your pyramid?', he would probably not have understood the question. But if he had been asked 'How long do you expect your building to last?', he would probably have replied 'For ever' and 4500 years can be considered a good approximation to 'for ever' in terms of the human life span.

Trial-and-error continued for many years to be the way in which manufactured objects were made more reliable. In the 19th century the steel plates of steam boilers were at first made thicker than was actually necessary (although the builders did not know this at the time). Over the years the thickness was reduced until catastrophic failures showed that the process had gone too far.

Attention to detail in design and emphasis on quality in manufacture were shown by Rolls Royce in the early years of this century to produce highly-reliable cars. But no coherent theory on how to make reliable objects was available.

Although a certain amount of work on aircraft reliability in the 1930s was of a theoretical nature, it was not until the German development of the V1 weapon in the 1939–1945 war that the basis of reliability theory was really laid. The V1 design team had great difficulty (fortunately for the British!) in making the weapon reliable. Only when a reliability assessment was made by Lusser was it realised that a system relying on many component parts was like a chain with many links. Since failure of any link causes failure of the whole chain it is thus necessary for each individual link to be highly reliable if the complete chain is to be reliable.

Immediately after the 1939–1945 war, little development of reliability theory took place. However, during the Korean war of 1950–1953 the USA found that the availability of its fighting aircraft was disastrously low, largely owing to the unreliability of the avionics, more and more of which was being carried by the aircraft. Thereafter the US military authorities made a determined and largely successful attempt to build a thorough understanding of reliability theory and practice.

The pace of understanding of reliability principles increased during the 1950s and 1960s. The next important milestone was the development, again by the military authorities, of the concept of System Effectiveness. This grew from the realisation that although reliability is an important system parameter, some wider method was needed for assessing the 'goodness' of several different competing systems.

The study of reliability is now well established. Journals dealing with reliability topics (*Microelectronics & Reliability* in the UK, *IEEE Transactions on Reliability* in the USA) are published regularly; many conferences are held, chiefly the Annual Symposium on Reliability & Maintainability in the USA and the biennial Reliability conference organised in the UK by the NCSR and IQA.

On the whole, system reliability can reasonably be considered to have shown steady improvement over the last 30 years, especially when the increased complexity of most systems is taken into account. Nevertheless the achievement of high reliability calls for never-ending labour and vigilance. Any relaxation of effort in this direction can be disastrous - as shown by the failure of the *Challenger* shuttle in 1986. The twin fields of Quality and Reliability (together with Maintainability) are likely to be those to which manufacturers of components or systems must pay increasing attention if they are to survive the competitive pressures of the 21st century. Increasing customer sophistication and awareness of the importance of reliability means that only those manufacturers who are dedicated to the highest product standards will survive in an increasingly competitive world.

1.3 Reliability and System Effectiveness

Reliability should be regarded as part of the overall study of System Effectiveness which has been defined as 'The probability that the system can successfully meet an operational demand within a given time when operated under specified conditions'. This definition is rather cumbersome; the essential point about System Effectiveness is that it attempts to measure the 'goodness' of a system and can thus be used to measure the effectiveness of competing systems especially when they perform the same end function in completely different ways. Reliability can be regarded as part of the total goodness of the system along with other major factors, such as Technical Performance.

Factors which are of importance in System Effectiveness usually include:

Technical Performance
Life Cycle Cost
Reliability
Size
Weight
Safety
Human Factors
EMC (Electro-Magnetic Compatability)
etc

The Technical Performance of a piece of equipment (or system) is always of prime importance since it is a measure of how well it fulfils its technical requirements. Nevertheless an equipment which performs only adequately but reliably may well be preferred to one which performs superbly but which often fails.

Life Cycle Cost (or Cost of Ownership) of equipment is of great importance to users since maintenance costs may be more significant than capital costs over the equipment's lifetime.* Thus the 'cost' of equipment should not be assessed by a prospective buyer purely on the capital cost but by the total costs which will be incurred during the equipment's lifetime (hence life cycle cost).

In System Effectiveness an attempt can be made to quantify (perhaps rather crudely) the 'goodness' of the important parameters so that an overall measure of 'goodness' may be obtained. Such a procedure is particularly useful when comparing different systems. Measures of System Effectiveness that have been used in the past include:

Performance × Reliability
Availability × Mission Reliability × Capability
Performance × Availability × Utilisation.

System Cost Effectiveness can be used as a Figure of Merit and is defined by:

System Cost Effectiveness = (System Effectiveness)/(Life Cycle Cost)

Reliability in one form or another (reliability, availability, dependability) appears in all these expressions and the reliability parameters are quantified by means of an analysis of the system. A glossary of terms used in Reliability Engineering and System Effectiveness is given in Appendix 4.

1.4 The Structure of this Book

This chapter is followed by a brief description of how systems are developed and the place of reliability in this development. This theme is taken up again in Chapter 16 on the Specification, Design, Manufacture and Installation of Reliable Systems.

Chapters 3 and 4 are devoted to the Statistics of Failure; the statistics are related to component failure, since it is the failure of one or more components which causes system failure. However, much of the statistical treatment is applicable to systems as well. The way that past failure statistics can be used to predict future failure patterns is described.

The failure of components is examined in Chapter 5 with reference to the effects of environment, power dissipation, complexity etc. on probability of failure. Sources of failure data are listed and a description is given of how they are used.

*Major users of telecommunication equipment may spend as much as 15 to 20 per cent of the capital cost of equipment *each year* on maintenance.

The simplest reliability system is a 'series' system in which the failure of any one component causes total system failure. The Series Reliability System is analysed in Chapter 6 and the way to predict failures on a probabalistic basis is described. The effects of repair are also considered.

Chapter 7 consists of a brief survey of methods of increasing system reliability and this serves as an introduction to the next two chapters on fault-tolerant systems. These are systems in which the failure of one, or possibly more components does not necessarily cause system failure.

The effect on a system of maintenance (corrective and preventive) in Chapter 10 is an expansion of the topic first introduced in Chapter 6.

Markov Analysis (Chapter 11) is a powerful method of system analysis based upon a diagram giving the various states which a system can take following various component failures. This introduces Chapter 12 which is devoted to Spare Parts Provisioning.

Software Reliability is a vast subject in its own right and only a brief discussion of the problem and some suggestions for alleviation of some of the problems associated with it are contained in Chapter 13.

The concept of maintenance (introduced in Chapters 6 and 10) is considered in more detail in Chapter 14 (on Maintainability). Chapter 15 (System Reliability Prediction) brings together the material of earlier chapters to show how system reliability is predicted in practice.

It is important for everyone engaged in engineering (buyer, vendor, user, maintenance engineer etc.) to understand at least some of the principles in Chapter 16 on the Specification, Design, Manufacture and Installation of Reliable Systems.

There is evidence from past data that initial deliveries of a new system design are unreliable, at least when compared with the reliability of later deliveries. Duane's method for predicting reliability growth is described in Chapter 17. System buyers are becoming increasingly sophisticated in demanding reliable equipment from manufacturers and Chapter 18 on Reliability Demonstration reviews methods of data analysis and shows how a test can be designed which is acceptable to both buyer and vendor.

Chapter 19 (Some Analytical Methods and Computer Software) considers briefly some reliability analytical methods which have not been pursued in detail in this book and lists commercially-available computer software. Chapter 20 (The Final Product) suggests that the reliability engineer (and all his colleagues!) should 'keep his eye on the ball' in the sense that ultimately he is seeking to play his part in the design, development, manufacture and maintenance of reliable products.

References in the text are given in the form Allen (1978); full details of each reference are provided in Chapter 21. A Bibliography is given in Chapter 22. A set of exercises is placed at the ends of appropriate chapters; answers and notes on the exercises are contained in Chapter 23. There are four Appendices: a set of Chi-squared (χ^2) Tables - Appendix 1; 5%, 95% and Median (50%) Rank Tables -

Appendix 2; the Elements of Probability Theory - Appendix 3; and a Glossary of Terms and Abbreviations - Appendix 4 Some exercises have been added to Appendix 3 and the answers to these are at the end of that appendix.

The emphasis throughout this book is on providing an insight into the principles and practice of reliability rather than on giving rigorous mathematical proofs, although references are given to more rigorous treatments which the interested reader can follow up The reader is assumed to understand simple differentiation and integration; some background knowledge of statistics would also be useful but is not essential. An asterisk* has been placed against some sections and exercises which require a greater knowledge (for example of Laplace Transform theory) than has generally been assumed; these can be omitted without losing the continuity of the argument. The reader is urged to attempt as many exercises as possible since these have been designed to provide a useful and instructive adjunct to the text.

Throughout the book an engineer is referred to as 'he'. This should not be taken to mean that the author has a prejudice against women engineers. (Reliability is a field in which women engineers can and do perform very well.) The usage is merely to avoid the tedious repetition of 'he or she'.

Exercises

1.1. List in order of importance the factors which you take into consideration when buying:
(a) a new car;
(b) a new colour television receiver.

1.2. A team of experts assessed the predicted technical performance and reliability (over the Design Life†) of three competing systems (A, B and C). For technical performance a score was awarded to each system out of a maximum of 10; reliability was predicted as a probability of surviving the Design Life. Results of these assessments are given in the table below. On the basis of this evidence, which system would you recommend for development?

System	*Technical performance*	*Reliability*
A	8/10	0.80
B	6/10	0.90
C	5/10	0.95

†See Glossary for definition.

2 Product Development

2.1 Introduction

It is suggested that good product development should follow the paths shown in Figures 2.1 and 2.2. At an early stage the Product Objectives are selected; these will consist of a number of parameters which are of prime importance for the system effectiveness of the new product and will together form an outline Product Specification. Apart from parameters relating to technical performance, there will almost certainly be some relating to cost (or life-cycle cost) and reliability. The flowchart shown in Figure 2.1 relates to the design phase before a model has yet been built. Reliability studies should be introduced as early as possible in this design phase since modifications introduced at later stages in the development are costly and time-consuming.

In the Development Phase (Figure 2.2) the Product Objectives should be refined in the light of the experience already gained and expanded to include other system effectiveness parameters (such as weight and size) which may not have been considered important enough to be included in the initial design phase.

An important part of each product development cycle is the system reliability prediction. This can be done from a knowledge of the system architecture and the reliabilities of the components from which the system is constructed.

2.2 Mission Profile and Design Life

The mission profile of a system (and of the components which comprise it) is a description of the duration and type of conditions which a typical system will encounter during its working life (or during a particular mission).

Three possible mission profiles are illustrated in Figure 2.3 for a telephone exchange, a guided missile and a telecommunication satellite. In each diagram the ordinate represents the system condition (storage, in operation etc.) while the abscissa represents time.

'Design life' is often used to describe the expected time that the system will be in use. This can range from 20 years for a telephone exchange to a few minutes for a guided missile.

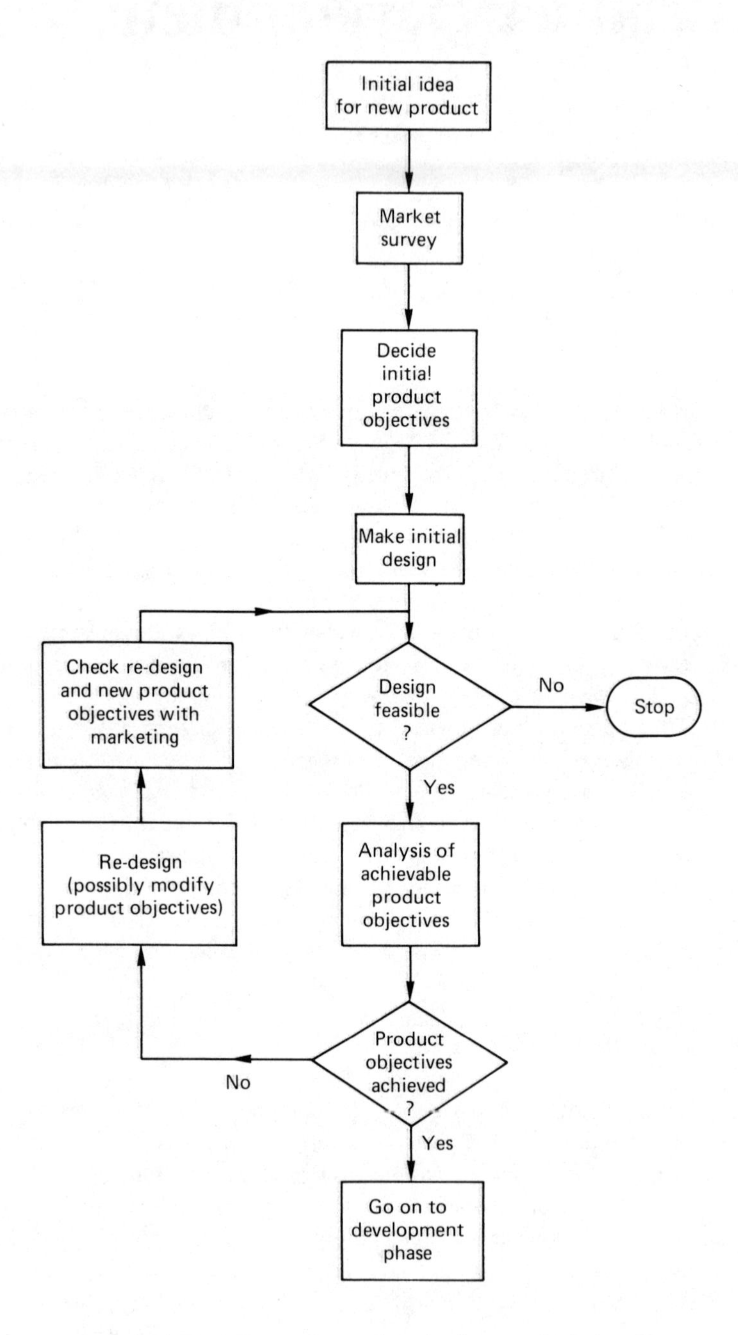

Figure 2.1 Flow diagram for product development in design phase

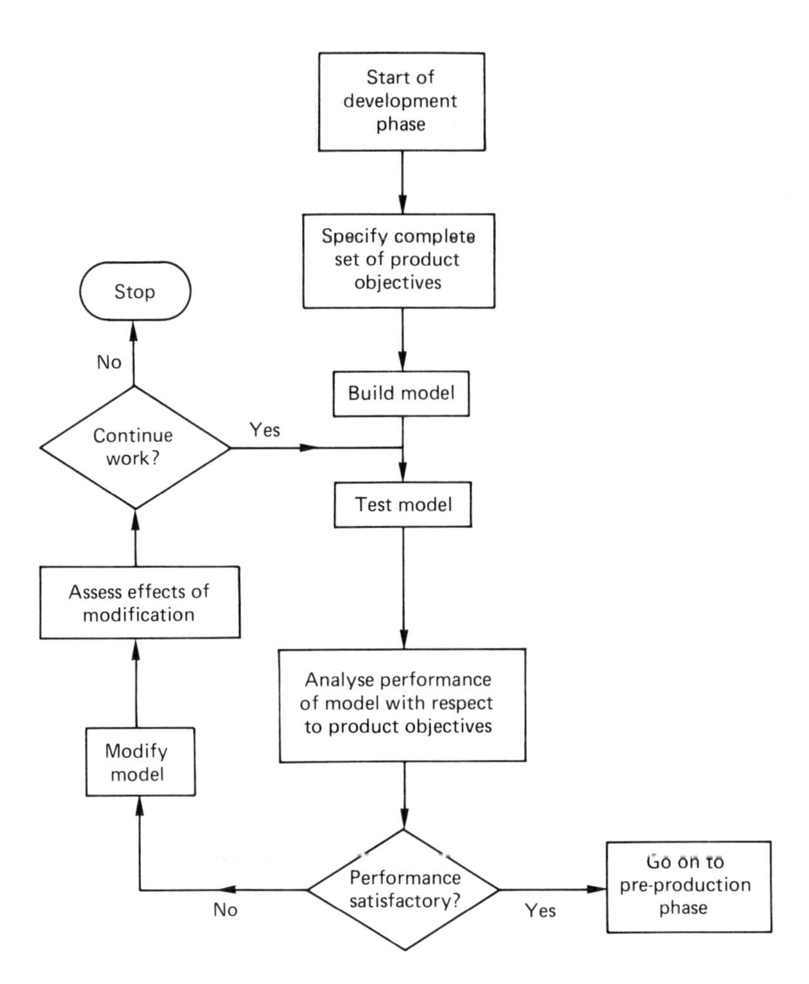

Figure 2.2 Flowchart for product development in development phase

Telephone Exchange

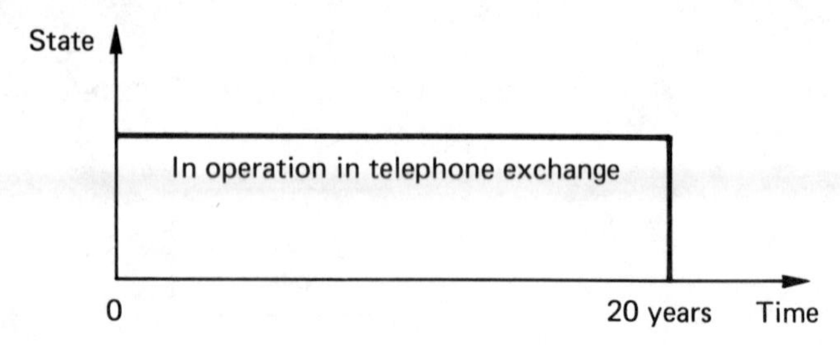

Guided Missile

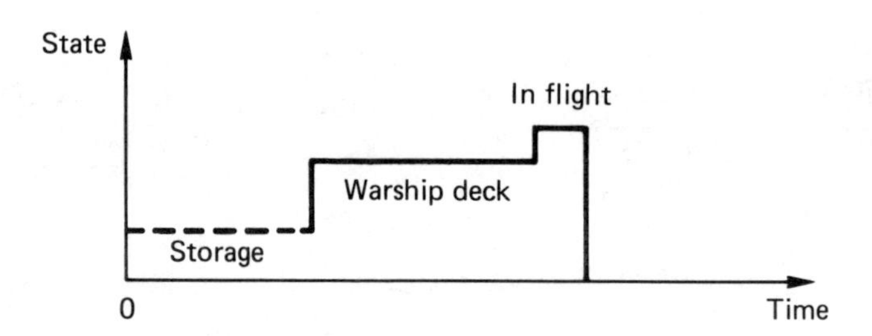

Telecommunication Satellite

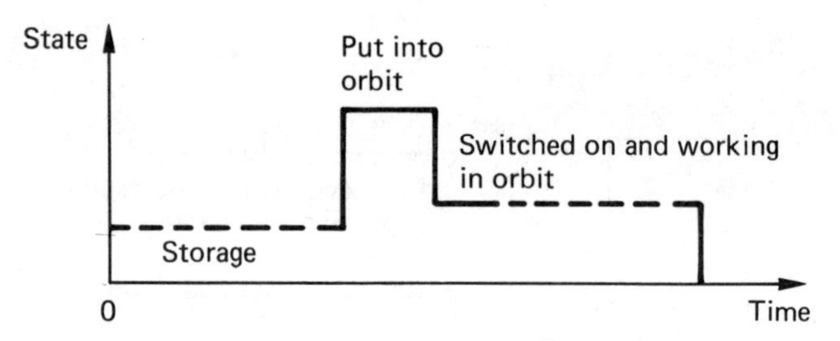

'Mission time' is often used to describe the expected time that the system is in use. This can range from 20 years for a telephone exchange to a few minutes for a guided missile

Figure 2.3 Mission profiles (the time axes are not drawn to scale)

Inside the overall mission profile, covering the design life of the system, there may be sub-profiles. This is the case for military aircraft which within their design lives will perform many short-term missions of interdiction, reconnaissance, ground attack etc. For other military systems the concept of a 'battlefield day' has been developed; this is a sub-profile of the conditions experienced by a military system during a typical day on the battlefield.

It is important for the system designer to know the mission profile because (as is shown below) reliability depends upon the system environment*; if the system is used in an environment for which it is not intended (for example, excessively high or low temperatures) then it will not be as reliable as it should be.

In a similar manner the design life may affect the choice of components which the system designer may use. A long design life implies that limited-life components (those which wear out during the design life) must be avoided if possible.

The mission profile and design life should form part of the system specification (see Chapter 16).

2.3 Reliability as an Ongoing Requirement

It takes a long time for a manufacturer to establish a good name for quality and reliability; but this good name can be lost overnight by neglect of an important Quality Assurance (QA) or reliability feature. It is thus important that reliability should not only be introduced early in new product planning, but that it should be an ongoing function. This means that systems and components should be regularly subjected to life testing and system reliability constantly monitored in the field. The disaster of the *Challenger* shuttle, which up to that time had had a good reliability record, was largely caused by the complacency of the management.

For success in the reliability field, dedication to Quality and Reliability is needed by managers, designers, manufacturers, installers and maintenance staff. Quality and Reliability may not be glamorous occupations, but the success of Japanese managers in changing their country's image from that of a producer of shoddy imitations to that of a producer of high-quality, reliable goods is an example of how dedication to Quality and Reliability can lead to oustanding commercial success.

It may be difficult to quantify the return on money spent on the Quality and Reliability functions, but refusal by management to allocate the necessary resources will only lead to economic decline. The near demise of all large-scale national UK car manufacturing is an example of such a commercial disaster.

*An interesting description of the environments in which a guided missile may have to operate is given in Short (1989).

2.4 Reliability and Safety (see also Chapter 16, section 16.4.8)

Tighter legislation on consumer protection has recently been introduced in several countries, notably the USA. This legislation chiefly affects OEMs (Original Equipment Manufacturers) but component manufacturers also need to consider the failures modes of their components to ensure that they may not be found culpably negligent by the courts With the tendency for component manufacturers to sell self-contained modules rather than single components, the need for failure mode analysis is increased.

2.5 Reliability and Economics

2.5.1 Balance between Cost and Reliability

The cost to the *producer* of a system can be expected to take the form shown in Figure 2.4(a). The corresponding curve of total cost for the *consumer* will be similar but probably displaced from that for the producer; the consumer's curve is shown in Figure 2.4(b). Thus, in general the OEM will wish to buy components of given reliability.

Although Figures 2.4(a) and 2.4(b) do represent the general form of variation of cost with reliability, they should be treated with some caution. In practice, neither the producer nor consumer will be able to make such an exact analysis of costs that he will be able to construct a cost/reliability curve in detail. It may also happen that the minima shown in the figures are not as sharp as the figures might imply.

High-reliability components are always likely to be expensive owing to the costly processes which are involved, but there may well be economic advantages to be gained in the manufacture of more ordinary components by attention to detail in the design and running of a product line.

2.5.2 Reliability Contracts

Contracts for the supply of equipment will usually contain reliability specifications (see Chapter 16). However, reliability cannot be checked by the buyer immediately upon delivery. To protect himself the buyer can:

(a) demand that the manufacturer gives warranties;
(b) obtain a promise to pay liquidated damages by the manufacturer if the reliability of the equipment in the field is not as high as he (the manufacturer) claims;
(c) require that the equipment pass a Demonstration Test (see Chapter 18);

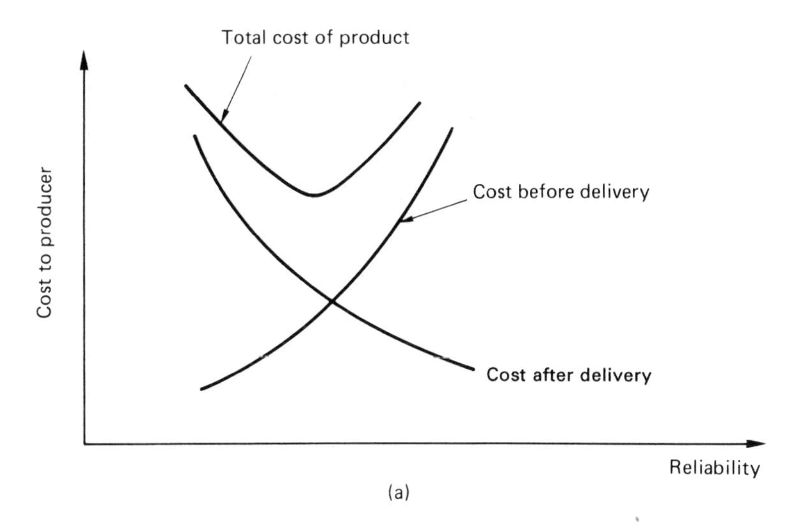

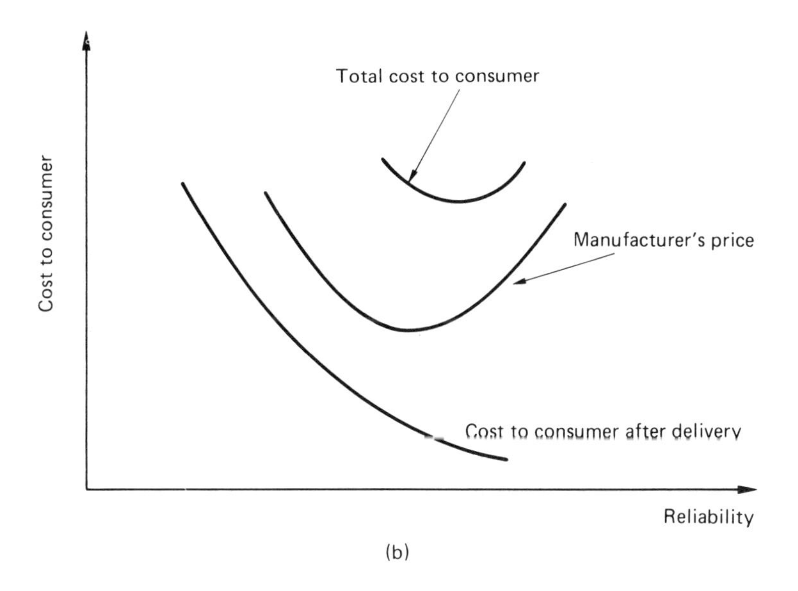

Figure 2.4 Balance between cost and reliability: (a) cost of reliability to producer; (b) cost of reliability to customer

(d) make some other arrangement, such as (i) agreeing to pay maintenance charges up to a specified maximum figure, above which the manufacturer must bear the full cost of maintenance or (ii) agreeing a reliability incentive contract (see O'Connor (1985)).

If the manufacturer accepts some of these conditions then he must take steps to assure himself that his equipment is good enough to prevent him from losing money (unless he is making such high profits that he can afford to pay damages for unreliability!).

To protect himself the manufacturer can:

(a) try to ensure that the components which he uses are as reliable as they should be;
(b) ensure that the equipment is properly manufactured;
(c) ensure that he understands the customer's (often badly stated) requirements;
(d) try to ensure that the equipment is properly installed, used and maintained;
(e) ensure that he understands the risks that he will run in any proposed demonstration test.

Both manufacturer and customer should:

(a) ensure that the contract is properly worded and that the definitions are clear (if necessary a method of arbitration should be agreed before the contract is signed);
(b) agree on a properly-designed statistical demonstration test.

(Either the manufacturer or the customer might still be unlucky, but at least they will have done the right thing!).

2.6 Reliability and Quality

There is a certain amount of overlapping between the studies of Reliability and Quality; how much overlapping depends upon the definitions which are used.

'Quality' has in the past been defined as 'conformance to specification'. Using this definition it seems quite clear that the quality function in a manufacturing organisation is responsible for ensuring that the goods delivered to a customer have measurable parameters (weight, size, surface finish etc.) which conform with the specification. Quality is thus concerned with the *immediate* satisfaction of the customer. This is different from the functions of the reliability group which is responsible for life-testing and predicting the failure characteristics of the goods, and in this way trying to ensure that the customer will be satisfied with the *future* performance of the goods.

Some quality practitioners now use the definition of quality as 'fitness for purpose'. This is a much less easily understood definition and would seem to include not only Reliability but Technical Performance and other considerations. Indeed it is more or less synonymous with System Effectiveness as used here. Throughout this book Quality will be defined as 'conformance to specification'.

2.7 Reliability Demonstration

The most convincing demonstration of reliability is a product which is recognised by users to be reliable over a considerable period of time. Rolls Royce Cars is a noteworthy example of a company which has gained and maintained a world-wide reputation for reliability. However since it may take time for a manufacturer to establish his reputation, or for a new system to prove itself reliable, customers may require manufacturers to mount a Demonstration Test which is a system trial statistically designed to show whether or not a product has an acceptable reliability. The design of Demonstration Tests is described in Chapter 18.

2.8 Budgeting for Reliability

During system development, a budget must be set for the various costs. It is not unreasonable for the cost of work on reliability to be 10 per cent of the total cost of development. It is difficult to extract a figure for the cost of software reliability alone since the writing of a program and its reliability are more intermingled than in the hardware case (see Chapter 13). It is not unusual for too little money to be allocated to reliability-related work initially, resulting in panic and cost over-run when the developed product is found to be unreliable.

2.9 Reliability and Management

It is essential that managements should understand the importance of reliability to its products (if they are manufacturers) or to its purchases (if they are consumers). For reliability engineers to do their jobs properly they must be provided with sufficient power to make their views be taken seriously. Far too many managements pay lip-service to reliability ('Yes, we do have a very good Reliability Department') without giving their engineers sufficient power and responsibility to take action when needed. In manufacturing companies the reliability department should have the power to prevent equipment deemed unreliable to be shipped out. This is a heavy responsibility, but the practice is widely used in Japan which now has a good reputation for reliability. In companies which purchase rather than manufacture systems, they too should have a department, however small, whose responsibility is to check the reliability (and quality) of goods being purchased.

Some companies combine quality and reliability together in one department. This is quite reasonable, so long as both parts can perform their separate tasks adequately and are each given sufficient power to ensure that their opinions are respected.

3 The Statistics of Failure – I

3.1 Introduction: Definition of Reliability

In this and the following chapter the statistics of failure are treated from the point of view of component failure although the statistical methods are also suitable for dealing with system failure data as well. Reliability is defined as

> 'the probability that a device or system will operate for a given period of time and under given operating conditions'*

It should be noted that because reliability is defined in this way, the mathematical methods of probability theory can be used in the development of reliability theory. It is not always easy to define what is meant by 'operate' and it is sometimes necessary to give a very precise definition of this term. For example a water tap might well be considered to be operational if, in a domestic setting, the 'off' position allowed one drop of water to escape every 10 seconds. But if the same rate of drippage occurred through a tap which was included in a piece of chemical engineering equipment, then this might be unacceptable and the tap would be considered to have failed. Thus the criteria which define the terms 'operate' and 'fail' would be different in the two cases even though the functions (and even perhaps the components) might be the same.

In this and the following two chapters, the failure of components is considered; 'failure' will mean catastrophic failure of a component (for example a resistor failing into a short- or open-circuit) or excessive parameter drift (so that the component is out of specification). From Chapter 6 onwards, system failure is considered; it is then necessary to define very precisely the meaning of 'failure mode' - see section 6.2.

*A more easily remembered (although less precise) definition is: 'Reliability is the probability of success as a function of time'.

3.2 Subjective and Objective Reliability Assessments

One may express an opinion on a piece of equipment based on one's own experience. ('Brand X television sets are very reliable.') Such assessments cannot be taken as objective since they depend so much upon an individual's experience.

However, if a sample of a manufacturer's output is taken and subjected to life-testing then, so long as it is a 'good' (that is representative) sample, objective statements about the reliability of the equipment can be made. But a prerequisite of a 'good' sample is that the manufacturer's output is reasonably homogeneous.

'Homogeneous' in this context means that the times to failure of the individual components all belong to a distribution which is reasonably consistent. For example, if the heights of men in the UK were measured, a histogram of the measurements would follow a reasonably smooth Gaussian (or Normal) distribution, and the 'population' would be considered to be homogeneous even though there will be some very short and some very tall men in the 'tails' of the distribution. If, however, there were a significant proportion of pygmies in the UK then the distribution of heights would show a 'hump' and close examination of the data would show two significantly different populations.

Where components are concerned, it is usually possible to consider that times to failure do form a homogeneous statistical population but only so long as component manufacturers maintain good quality assurance procedures. Reliability theory can be regarded as a three-legged stool which uses as its three supporting legs the disciplines of statistics, probability theory and (electrical and mechanical) engineering.

3.3 The Mechanism of Failure

Failure of a component may be caused by bad workmanship or by over-stressing of a component or by wear-out. But components which are well-made, not over-stressed and not worn-out may still fail through a sudden accumulation of internal stresses. Such failures occur randomly in time and give rise to a certain probability of failure even for well-made components.

The reason why random failures occur may be shown pictorially as in Figure 3.1 (adapted from Bazovsky (1961)). This shows the internal stresses in a component as a function of time when a constant external stress is applied. The internal stress fluctuates randomly, and if a peak rises above the maximum stress that the component can withstand then it fails. If a stronger component is used then (with the same external applied stress) the sudden failures will occur less frequently since a higher random peak is needed before the maximum stress line is crossed. When a high-quality component is used it is to be expected that the high stress peaks will be less numerous than those for the low-quality component of the same rating, and that the high-quality component will on average last longer than the low-quality component.

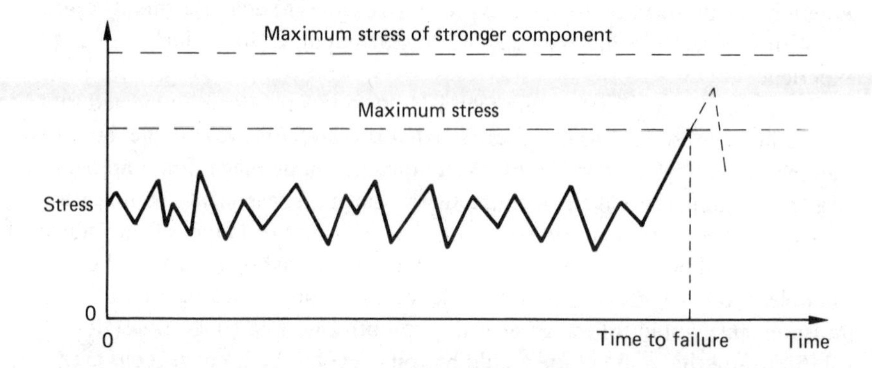

Figure 3.1 Pictorial representation of the internal stresses in a component as a function of time [Igor Bazovsky, ***Reliability Theory and Practice***, © 1961, p. 147. Adapted by permission of Prentice-Hall, Englewood Cliffs, New Jersey]

There is a close analogy between the failure of well-made components and the radio-active decay of some types of atom. In both cases it is not possible to specify when a certain individual will fail (or decay), but it is possible to make reasonably accurate statistical predictions about how a whole population of nominally identical individuals will fail (or decay). The analogy can be taken further in that the form of the failure (decay) curve is often exponential. In reliability theory this is expressed in terms of a constant failure rate; in atomic physics an equivalent half-life is usually quoted.

3.4 Relevant and Non-relevant Failures

In assessing statistical life-test data on components or systems it is important to segregate the relevant from the non-relevant failures since otherwise unduly pessimistic conclusions will be drawn.

Some reasons for the occurrence of non-relevant failures are:

Faulty preventive maintenance } (see Chapter 10)
Faulty corrective maintenance }
Misuse
An external failure
Faulty connection
System design inadequacy

It is also surprising how often a repairman reports 'fault not found' when called to repair an alleged failure.

3.5 The Reliability Function $R(t)$

If a large number (n) of components are put under test for an indefinite period of time, so that all eventually fail, the successive times-to-failures, $t_1, t_2, t_3, \ldots,$ t_n, may be plotted as shown in Figure 3.2.

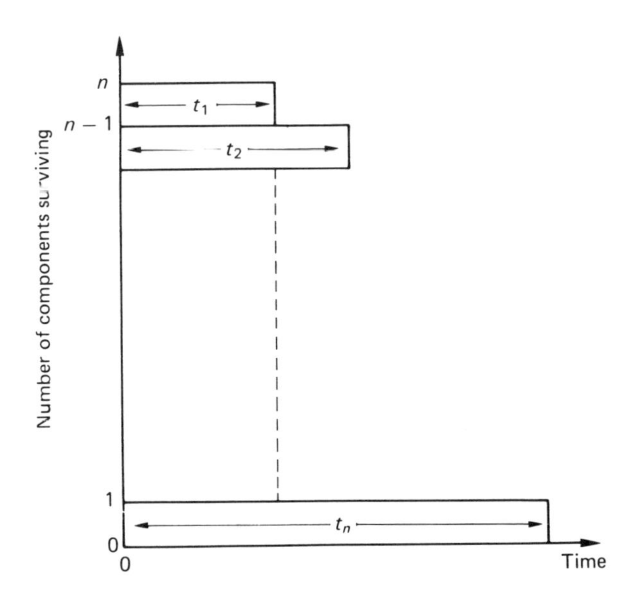

Figure 3.2 Bar chart showing times to failure (the horizontal and vertical scales are both linear)

If we normalise the ordinates by dividing by n then we obtain a graph which starts from unity at time $t = 0$ whatever the value of n we use, as shown in Figure 3.3; the ordinate now becomes the fraction of components surviving at time t.

The height of each bar is now $1/n$ and so the mean time to failure (MTTF) is

$$\frac{t_1 + t_2 + t_3 + \ldots + t_n}{n}$$

$$= \frac{t_1}{n} + \frac{t_2}{n} + \frac{t_3}{n} + \ldots + \frac{t_n}{n}$$

= area under graph

If these results are generalised by allowing n to become extremely large then we obtain a smooth curve. The function being plotted is called the reliability (or survivor) function $R(t)$; its general shape is shown in Figure 3.4. Since components are not replaced as they fail, the slope of $R(t)$ must always be negative (or more accurately, non-positive).

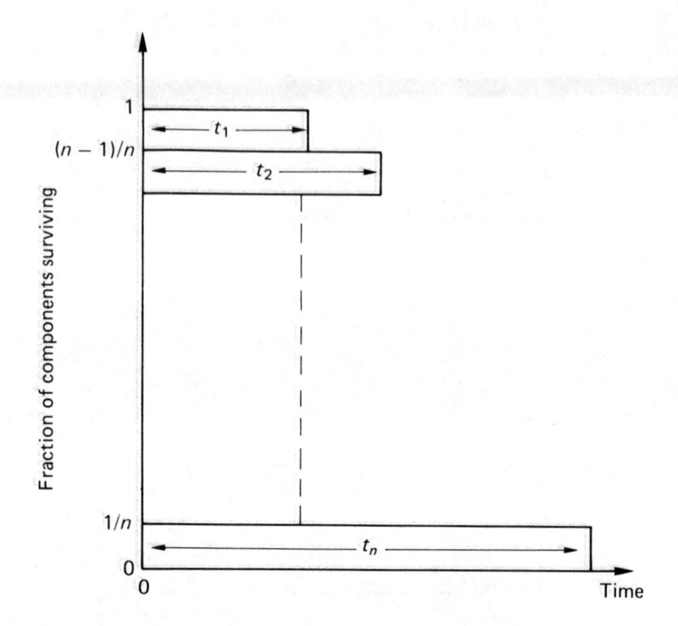

Figure 3.3 A normalised version of Figure 3.2 (the horizontal and vertical axes are both linear)

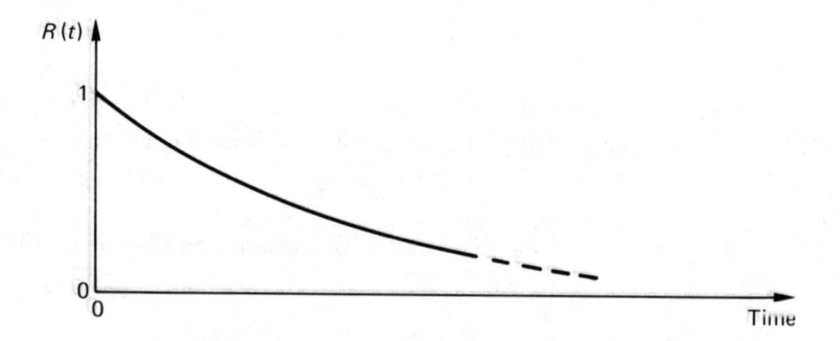

Figure 3.4 The function $R(t)$ plotted against time

By definition $R(0) = 1$ and we take it as axiomatic that eventually all the components will fail so that $R(\infty) = 0$. By generalising the above result for the MTTF(θ) we can put*

*A rigorous proof of this expression is not as easy as it might appear. The outline of a formal proof appears in Evans (1969).

$$\theta = \int_0^\infty R(t)\,\mathrm{d}t$$

If the components are homogeneous we can consider the $R(t)$ curve to represent the reliability of a single component taken at random from the population. We assume that all the components are functioning at $t = 0$ and this corresponds to $R(0) = 1$ that is, the probability that our randomly selected component is functioning at $t = 0$ is unity. (A probability of unity corresponds to certainty of survival.) On the other hand, we have taken it as axiomatic that all the components fail as $t \to \infty$ and so $R(\infty) = 0$ corresponds to our certainty of failure for our randomly selected component. Between $t = 0$ and $t = \infty$, $R(t)$ will have a value between unity and zero corresponding to the probability of our randomly selected component surviving to time t.

Thus, having plotted the reliability function for a large number of components we can now use this curve to predict the probability of survival of a randomly selected component so long as it comes from the same statistical population as those tested and that it is also subjected to the same stress levels and ambient conditions.

The reliability function may thus be defined as the probability of survival of a single component (or system) as a function of time.*

It is necessary for the reader to understand the key nature of the process described above. We have taken past data and, on the assumption that future (untested) components will be statistically identical with those already tested, used this as a basis for future prediction. As a symbol of the process involved, Figure 3.5 shows the Roman god Janus who looks both backwards and forwards.

Figure 3.5 The Roman god Janus who looks both backwards and forwards

The danger of this prediction method is that the components which are built into the systems may not be statistically identical to those that have been tested. It requires only a small misadjustment to a machine or a slightly contaminated batch of raw materials to make this assumption invalid. Thus for us to have any

*The IEC definition, 'The reliability function $R(t)$ is the probability that a component (or system) can perform a required function under stated conditions for a given period of time', is very clumsy.

confidence in our predictions (for both components and systems) it is vital that the component manufacturer should maintain a high level of quality assurance at all stages of component manufacture.

The mean time to failure, θ, is often used as a criterion of the goodness of a component; but it should be noted that the percentage of components which will survive until θ depends upon the shape of the reliability function. If the reliability function is shaped as $\exp(-\lambda t)$ then only 37 per cent (approximately) of the components will survive to θ (see below).

3.6 The Failure Function $F(t)$

This is defined as the complement of $R(t)$:

$$F(t) = 1 - R(t)$$

Thus by definition:

$$F(0) = 0$$

$$F(\infty) = 1$$

3.7 The Failure Density Function $f(t)$

By definition:

$$f(t) = -\mathrm{d}R/\mathrm{d}t$$

Since the slope of $R(t)$ is always negative (non-positive), it follows that $f(t)$ is always positive (non-negative).

In fact, $f(t)$ is a type of function known in statistics as a 'probability density function' since

(a) it is always positive or zero

(b) $\displaystyle\int_{t_1}^{t_2} f(t)\,\mathrm{d}t = R(t_1) - R(t_2)$

= proportion of failures occurring in the interval t_1 to t_2

(c) $\displaystyle\int_0^{\infty} f(t)\,\mathrm{d}t = R(0) - R(\infty) = 1$

An alternative approach to the derivation of the MTTF is to consider that it is the first moment of the failure density function:

$$\text{MTTF} = \int_0^{\infty} t\,f(t)\,\mathrm{d}t$$

This is equivalent to the previous definition since on integrating by parts:

$$\int_0^\infty t\,f(t)\,\mathrm{d}t = [t\,R(t)]_0^\infty + \int_0^\infty R(t)\,\mathrm{d}t = \int_0^\infty R(t)\,\mathrm{d}t$$

so long as $tR(t) \to 0$ as $t \to \infty$. This condition will hold for all the reliability functions considered here but it is possible to construct a function for which the condition does not hold even though it has the other desired properties (see Evans (1969)).

3.8 The Failure Rate $\lambda(t)$

By definition:

$$\lambda(t) = -\frac{(\mathrm{d}R/\mathrm{d}t)}{R(t)} = \frac{f(t)}{R(t)}$$

By considering a small but finite interval of time Δt we can see that

$$\lambda(t)\,\Delta t \approx \frac{\Delta R}{R}$$

and this may be identified as the probability that a certain component will fail in the interval of time t to $(t + \Delta t)$, given that it has survived up to the time t. The failure rate can thus be regarded as the percentage of components working at time t which have failed in unit time (Δt). Failure rate has the dimensions of 1/(time) and it is often quoted in units of 10^{-9} per hour. These units are called fits.

The relationship between $\lambda(t)$ and $R(t)$ is obtained as follows:

$$\lambda(t) = \left(-\frac{\mathrm{d}R}{\mathrm{d}t}\right)\Big/R$$

so that $\mathrm{d}R/R = -\lambda(t)\,\mathrm{d}t$

Integrating the left- and right-hand sides of this equation from 1 to R and 0 to t respectively gives

$$\ln(R) = -\int_0^t \lambda(u)\,\mathrm{d}u$$

where u has been substituted for t as a dummy variable. This equation yields

$$R(t) = \exp\left[-\int_0^t \lambda(u)\,\mathrm{d}u\right]$$

Other terms used for the function defined here as 'failure rate' are 'hazard rate', 'age specific failure rate' and 'force of mortality'. However, definitions may vary and the reader is advised to check each author's usage of these terms.

3.9 Inter-relationships between $R(t)$, $f(t)$ and $\lambda(t)$

The relationships between $R(t)$, $F(t)$ and $f(t)$ are shown in Figure 3.6. Since $R(t)$ is the probability of survival from time 0 to t, its value must be represented by the area to the right of t in Figure 3.6. Correspondingly, $F(t)$ is the probability of failure in time 0 to t and is represented by the area to the left of t in Figure 3.6.

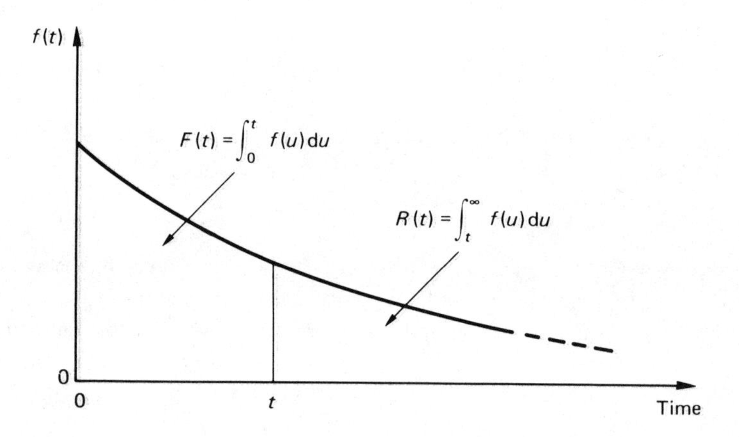

Figure 3.6 Illustration of the relationship between $R(t)$, $F(t)$ and $f(t)$

If any one of $R(t)$, $f(t)$ or $\lambda(t)$ is known (as a mathematical function) then the other two (and hence $F(t)$) can be derived from it. Thus these functions are not independent.

If $R(t)$ is known then

$$f(t) = -\mathrm{d}R/\mathrm{d}t \quad \text{and} \quad \lambda(t) = f(t)/R(t)$$

If $f(t)$ is known then

$$R(t) = \int_t^\infty f(u)\,\mathrm{d}u \quad \text{and} \quad \lambda(t) = f(t)/R(t)$$

If $\lambda(t)$ is known then

$$R(t) = \exp\left[-\int_0^t \lambda(u)\,\mathrm{d}u\right] \quad \text{and} \quad f(t) = -\mathrm{d}R/\mathrm{d}t$$

3 10 General Form of Failure Rate (the 'Bathtub' Curve)

The most general form of the failure rate $\lambda(t)$ as a function of time is shown in Figure 3.7; it is often known as the 'bathtub' curve because of its shape. It may roughly be divided into three portions: a period of 'early' (or infant) failures where the failure rate starts at a high value and falls rapidly; a period when the failure rate is approximately constant; and a 'wear-out' period when the failure rate rises rapidly again.

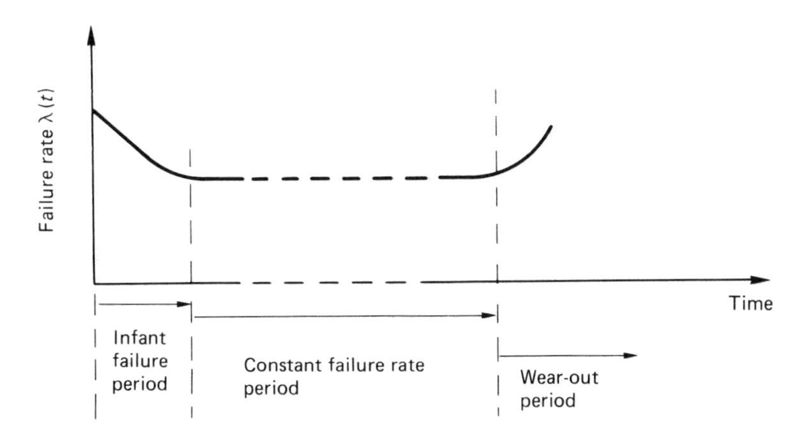

Figure 3.7 The 'bathtub' curve

It should be understood that the designation of the time of changeover from one period to another is fairly arbitrary and that overlapping of the types of failure occurs. Thus the last of the 'early failures' will occur during the 'constant failure rate period' as also will the first of the 'wear-out' failures. Similarly, stress-related failures of the type which predominate in the 'constant failure rate period' will occur in the 'early failure' and 'wear-out' periods.

The Useful Life of a component is the period of time from the commencement of use to that time at which wear-out causes a significant increase in the failure rate.

3.11 Constant Failure Rate (CFR)

For many electronic components it is possible to consider the failure rate to be constant. This is a very useful concept, and has important consequences (see below). The graph of $\lambda(t)$ becomes a straight line which continues to infinity as shown in Figure 3.8.

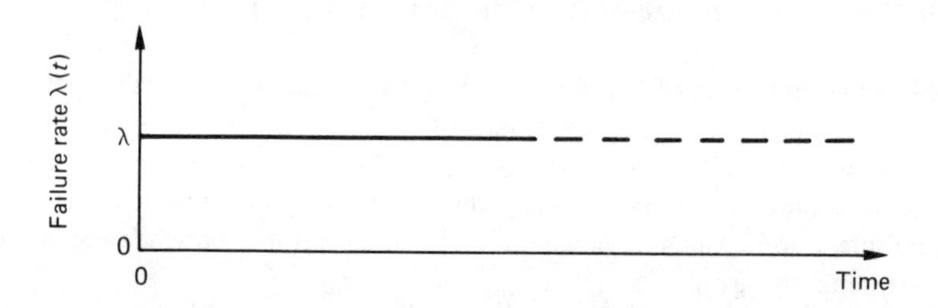

Figure 3.8 Constant failure rate against time

In this book when $\lambda(t)$ is constant, it is designated by the symbol λ; the nomenclature is not uniform and care should be taken in interpreting each author's symbols for failure rate, MTTF, etc.

The effect of a constant failure rate is that

$$-\frac{(\mathrm{d}R/\mathrm{d}t)}{R} = \lambda$$

Re-arranging and integrating this expression between $t = 0$, $R = 1$, and the dummy limits t' and R' gives:

$$\int_1^{R'} \frac{\mathrm{d}R}{R} = -\int_0^{t'} \lambda \,\mathrm{d}t$$

whence

$$[\ln(R)]_1^{R'} = -[\lambda t]_0^{t'}$$

and

$$\ln(R') = -\lambda t'$$

or

$$R' = \exp(-\lambda t')$$

The dashes are only used to denote the limits of integration and may now be discarded giving

$$R(t) = \exp(-\lambda t)$$

but only when the failure rate is constant.

It should be noted that this function does satisfy the requirements that $R(0) = 1$ and $R(\infty) = 0$; its shape is shown in Figure 3.9.

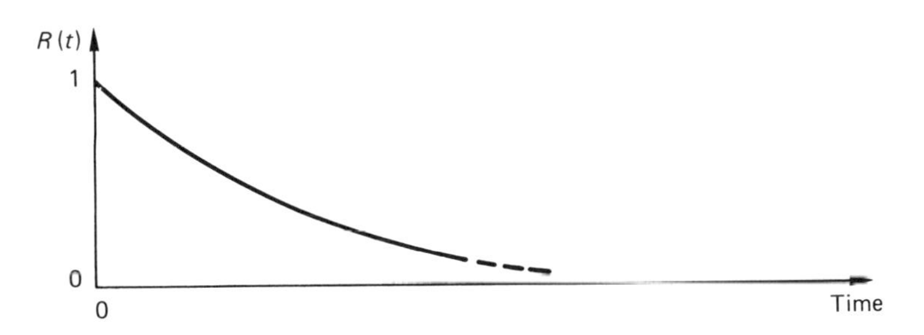

Figure 3.9 The function $\exp(-\lambda t)$

3.12 Calculation of MTTF (θ) when the Failure Rate is Constant

3.12.1 Without Wear-out

If the component never wears out then from the above:

$$\theta = \int_0^\infty R(t)\,dt = \int_0^\infty \exp(-\lambda t)\,dt$$

$$= [-(1/\lambda)\exp(-\lambda t)]_0^\infty = 1/\lambda$$

Thus, when the failure rate is constant:

$$\text{MTTF} = \theta = 1/(\text{failure rate})$$

Fraction of components surviving to time $\theta = 1/e \approx 0.37$.

3.12.2 With Wear-out

The failure rate curve for a CFR component with wear-out may be shown in an idealised form as in Figure 3.10(a) in which the failure rate becomes infinite at time $t = t_w$. This corresponds to putting:

$$R(t) = \exp(-\lambda t) \quad 0 \leqslant t \leqslant t_w$$

$$R(t) = 0 \quad t > t_w$$

Thus

$$\theta = \int_0^\infty R(t)\,dt = \int_0^{t_w} \exp(-\lambda t)\,dt$$

$$= [-\exp(-\lambda t)/\lambda]_0^{t_w} = [1 - \exp(-\lambda t_w)]/\lambda$$

Two extreme cases can now be considered:

(a) If $1/\lambda \ll t_w$, that is $1 \ll \lambda t_w$, then
$\theta \approx 1/\lambda$ as illustrated in Figure 3.10(b).
(b) If $1/\lambda \gg t_w$, that is, $1 \gg \lambda t_w$, then

$$\exp(-\lambda t_w) \approx 1 - \lambda t_w$$

and $\theta \approx t_w$ as illustrated in Figure 3.10(c).

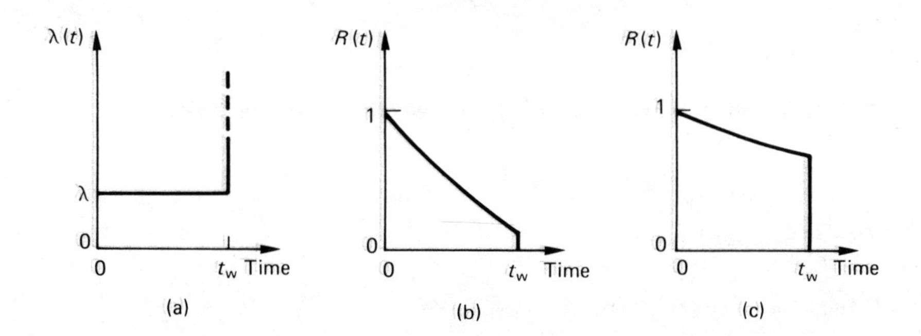

Figure 3.10 Bathtub curve with wear-out and two possible forms of $R(t)$

In cases such as those illustrated by Figure 3.10(c), the practice of specifying $1/\lambda$ as 'the MTBF' is to be deprecated. The component should be described as having 'a failure rate of λ during its useful life and a wear-out time of t_w'.

3.13 Summary of the Mathematical Functions Applicable to Constant Failure Rate Components

$$R(t) = \exp(-\lambda t)$$

$$F(t) = 1 - \exp(-\lambda t)$$

$$f(t) = \lambda \exp(-\lambda t)$$

$$\lambda(t) = \lambda$$

3.14 Memoryless Property of the Exponential Distribution

Consider that a number of components are put on test (without replacement) and it is found that they fail exponentially. If the test is stopped at time t_1 and the non-failed components are taken away and put on another test, then the second test timed over $t' = t - t_1$ will also appear to have an exponential shape. The number surviving to time t is $n(t) = n(0)\exp(-\lambda t)$ so that

$$R(t) = \frac{n(t)}{n(0)} = \exp(-\lambda t)$$

for the first test and

$$R(t') = \frac{n(t)}{n(t_1)} = \frac{\exp(-\lambda t)}{\exp(-\lambda t_1)} = \exp[-\lambda(t - t_1)] = \exp(-\lambda t')$$

for the second test. This is shown diagrammatically in Figure 3.11.

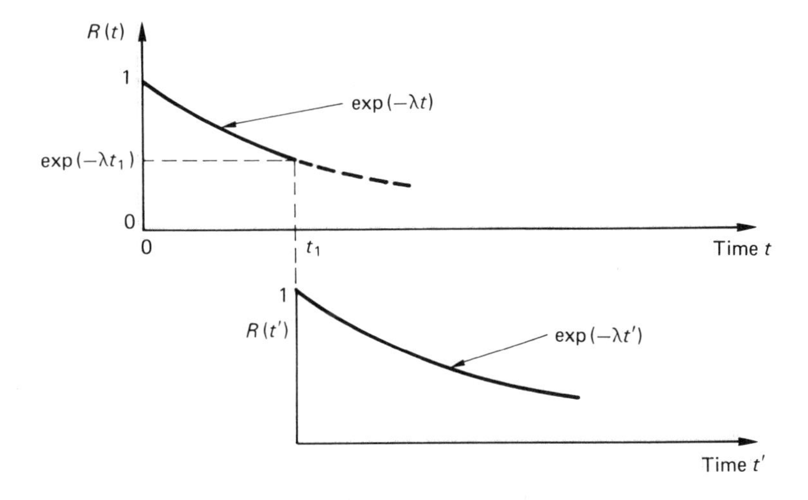

Figure 3.11 Illustration of the memoryless property of the exponential distribution

Thus it follows that if a constant failure rate component has survived up to time t_1 then it can be considered to be 'as good as new' at time t_1 and its reliability function will again be of exponential form; thus the life expectation (the mean residual life) remains constant whatever the age of the component. This 'memoryless property' holds only for constant failure rate components* (and also for constant failure rate systems: see Chapter 6).

*The exponential distribution is the only continuous distribution with this property. See Leemis (1988).

3.15 The Mean Residual Life

The Mean Residual Life, $L(t)$ is the expected life of a component which has already survived a time t; the time already elapsed from 0 to t is not counted as part of $L(t)$. It is in general a function of t although, as shown above, it is constant and equal to $1/\lambda$ in the special case of a CFR component. By analogy with the method of calculating the MTTF, $L(t)$ can be found from

$$L(t) = \frac{\int_t^{\infty} R(u)\,\mathrm{d}u}{R(t)}$$

where $R(t)$ acts as a 'normalising' factor which makes $R(u)/R(t)$ equal to unity at time $u = t$.

Clearly $L(0) = \text{MTTF}$ for any distribution of times to failure.

3.16 Expected Number of Failures in a Given Time

3.16.1 Renewal

Consider that a component is placed on life-test under given ambient and stress conditions. When it fails it is replaced instantaneously by a similar component or system. (In practice, the replacement will not be instantaneous but an equivalent result is obtained by considering the clock which times the test to be stopped at the instant of failure and re-started when the new component is switched on.) This process is continued until the test is halted. The treatment of data obtained in this way is the province of renewal theory. (Failure followed by repair may be treated as an alternating renewal process.)

The data generated in this test may be displayed as in Figure 3.12(a) in which t_i is the length of time which the ith component survives and a cross represents the time that a failure takes place. The test is halted at time T. (This is then a time-truncated test as described below in section 4.8.3.)

It may happen that no failure takes place in the time 0 to T. If $f(t)$ is the failure density of the device then since $\int_0^T f(t)\,\mathrm{d}t$ is the probability that a failure has occurred in the interval 0 to T, it follows that $\int_T^{\infty} f(t)\,\mathrm{d}t$ must be the probability that no failure has occurred in the same interval. We can put

$$\Pr(0) = \int_T^{\infty} f(t)\,\mathrm{d}t$$

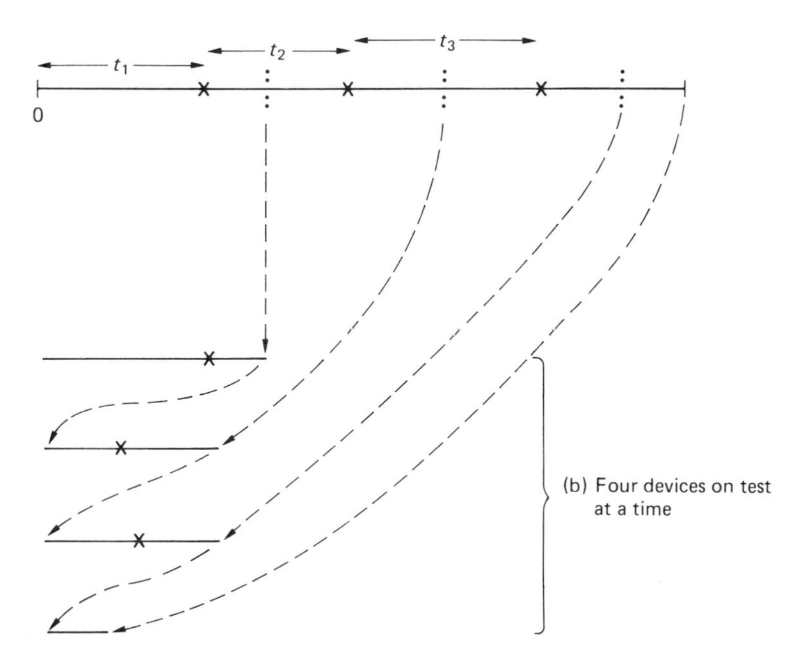

Figure 3.12 Failure data

*3.16.2 Convolution of Failure Density Functions

Now consider the two-fold convolution of $f(t)$ represented by $f(t) * f(t)$. This represents a new failure density which we shall denote by $f(t)^{(2)}$ so that $\int_0^T f(t)^{(2)} \mathrm{d}t$ represents the probability that two (or more) failures have occurred in the interval 0 to T. It follows that $\int_T^\infty f(t)^{(2)} \mathrm{d}t$ represents the probability that 0 or one failure has occurred in this interval.

Thus the probability that exactly one failure has occurred in the interval 0 to T is

$$\Pr(1) = \int_T^\infty f(t)^{(2)} \mathrm{d}t - \int_T^\infty f(t) \mathrm{d}t = \int_T^\infty [f(t)^{(2)} - f(t)] \, \mathrm{d}t$$

The above argument may be extended to give

$$\Pr(i) = \int_T^\infty [f(t)^{(i+1)} - f(t)^{(i)}] \, \mathrm{d}t$$

where $f(t)^{(i)}$ represents the convolution of $f(t)$ with itself i times.

The derivation of the probabilities Pr(0), Pr(1) . . . is illustrated in Figure 3.13. The function $f(t)$ is taken to be $\exp(-\lambda t)$, corresponding to the constant failure rate case, although the above treatment is correct for any valid form of $f(t)$.

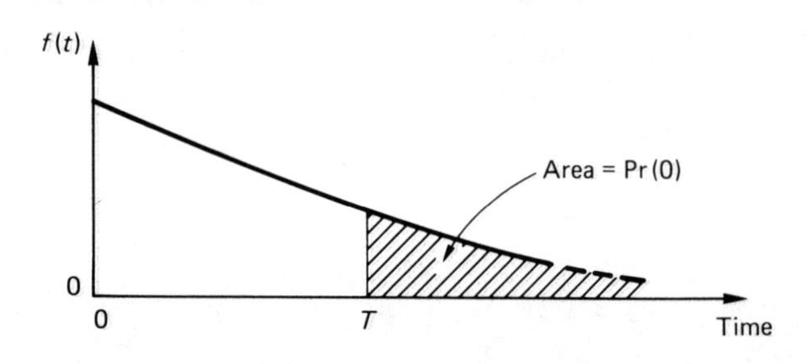

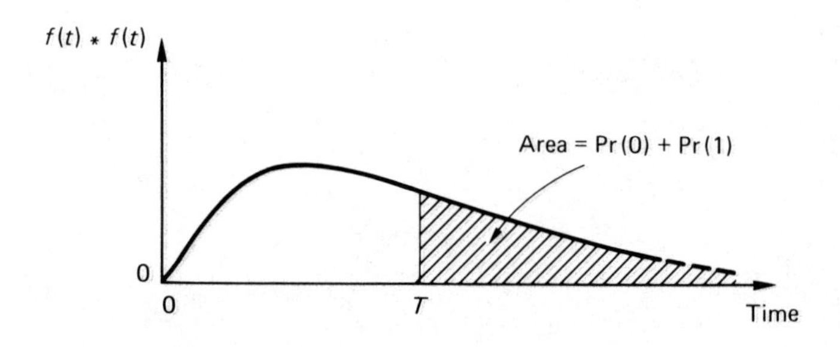

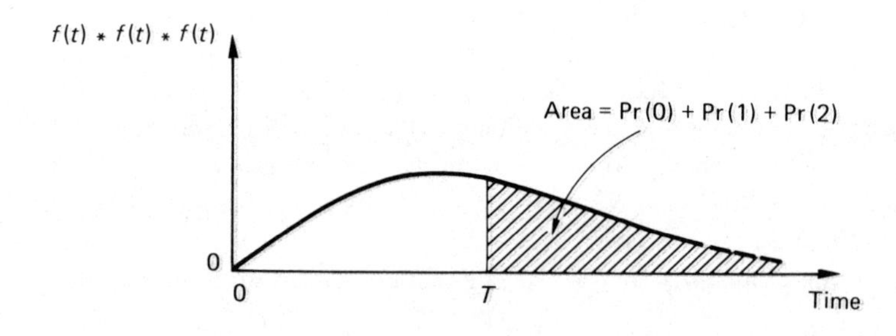

Figure 3.13 Graphical illustration of the calculation of Pr(0), Pr(1) and Pr(2)

*3.16.3 Convolution in the CFR Case

The convolved functions are most easily calculated using the Laplace transformation. Using L to denote this transformation, then if

$$f(t) = \lambda \exp(-\lambda t)$$

$$Lf(t) = \lambda/(s + \lambda)$$

and $$Lf(t)^{(i)} = [\lambda/(s + \lambda)]^i$$

so that

$$f(t)^{(i)} = L^{-1}[\lambda/(s + \lambda)]^i = \frac{\lambda^i t^{i-1} \exp(-\lambda t)}{(i-1)!}$$

It follows that $\Pr(i)$, the probability of i failures, is given by:

$$\Pr(i) = \int_T^\infty \frac{\lambda^{i+1}\, t^i \exp(-\lambda t)}{i!}\, dt - \int_T^\infty \frac{\lambda^i\, t^{i-1} \exp(-\lambda t)}{(i-1)!}\, dt$$

$$= \frac{(\lambda T)^i \exp(-\lambda T)}{i!}$$

The last result is obtained by taking the first integral in the previous line and integrating by parts.

3.16.4 Expected Number of Failures – CFR Case

It has been shown in the previous two sections that in the CFR case the failure probabilities for a component with CFR λ in T component hours are given by the terms of a Poisson series with parameter λT. That is

$$\Pr(0) = \exp(-\lambda T)$$

$$\Pr(1) = (\lambda T) \exp(-\lambda T)$$

$$\Pr(2) = (\lambda T)^2 \exp(-\lambda T)/2!$$

$$\vdots \qquad \vdots$$

$$\Pr(i) = (\lambda T)^i \exp(-\lambda T)/i!$$

etc.

So far, the derivation of the $\Pr(i)$ has been based upon a test where only one device is under test at any time, as shown in Figure 3.12(a). However, owing to the memoryless property of a constant failure rate device, the data in Figure 3.12(a) is clearly equivalent to that shown in Figure 3.12(b), so long as the total device hours (T) and the number of failures (k) are the same in both cases. The picture of Figure 3.12(b) can obviously be generalised to represent any number of (constant failure rate) components generating T component hours in a completely haphazard fashion; the probabilities of 0, 1, 2, . . . failures are the terms of the Poisson series.

The expected number of failures in a given period T, irrespective of the number of components used in generating T, is the mean of the Poisson distribution with parameter λT. This mean is just λT; the result could be expected intuitively since it represents T/(mean time to failure).

Returning to the convolution of $f(t)$, when $i = 2$:

$$f(t)^{(2)} = \lambda^2 t \exp(-\lambda t)$$

so that reliability corresponding to this failure density is

$$R_s(t) = \int_t^{\infty} f(u)^{(2)} du = \lambda t \exp(-\lambda t) + \exp(-\lambda t)$$

This result is used in section 8.4.

3.16.5 Non-constant Failure Rate Case

This case is more difficult to deal with than the constant failure rate case since only the latter has the memoryless property. If a single component is put on test and replaced (renewed) immediately upon failure, then over a given time T, the expected number of failures, is again T/(mean time to failure). Further treatment of this problem is given in textbooks on renewal theory, such as Cox (1962).

3.17 Composite Failure Pattern

3.17.1 Constant Failure Rate Case

Consider a group of n identical constant failure rate components with failure rate λ. If they are all put on test under identical conditions and immediately replaced (renewed) upon failure, then the composite pattern of failures in the group is (statistically) identical with that of one component with constant failure rate $n\lambda$. The form of the composite pattern is illustrated in Figure 3.14 in which $n = 4$.

The composite* distribution of the failures is exponential (see Cox (1962)) and hence the expected number of failures in a total of T component hours is $T/n\lambda$; the probabilities of 0, 1 2, . . . failures in time T are the terms of the Poisson series with parameter $n\lambda T$. Moreover, as a result of the memoryless

*In Renewal Theory the process shown in Figure 3.14 is called 'superposition' and the data resulting from it are called 'pooled' data.

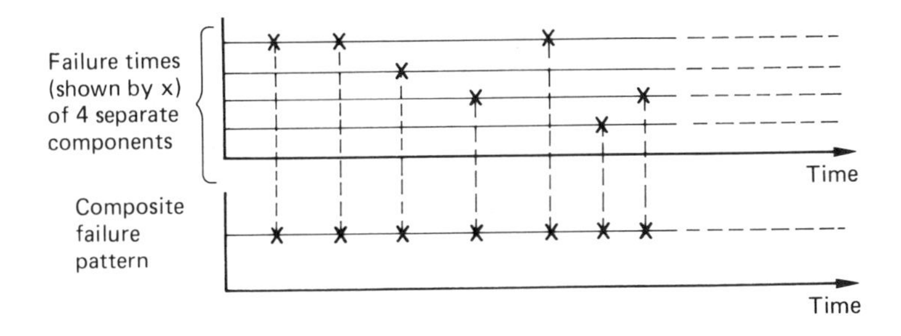

Figure 3.14 Illustration of the composite failure pattern of four identical components

property of the exponential distribution, the time from which the test may be considered to start is perfectly arbitrary.

3.17.2 Non-constant Failure Rate Case

This case is more difficult to deal with; it is treated in textbooks on renewal theory, see for example, Cox (1962).

A general result is that for any valid failure distribution (except the Dirac delta distribution) failures will, after a sufficiently long period, become exponentially distributed and hence the failure rate is constant and equal to n/(mean time to failure on one component). The results of a computer simulation of this problem and comments on the length of the 'settling-down' period are given in Grosh and Lyon (1975).

3.18 A Note on the Meaning of Randomness

It is often assumed (incorrectly) that 'randomness' implies that data are compatible with the exponential distribution only. This is not so; randomness (in statistical usage) implies only that an exact outcome is unpredictable. Thus times to failure which might be characterised by the exponential, log-normal or rectangular failure density function are all 'random'.

Exercises

3.1. Convert $\lambda = 0.1$ per cent/1000 hours into an equivalent failure rate in fits.
Convert $\lambda = 1000$ fits into an equivalent failure rate in per cent/100 hours.

3.2. Consider 10 000 components placed on life test. If the failure rate is constant and equal to 1 per cent/hour, plot the number of components that you would expect to be surviving as a function of the time t from $t = 0$ to $t = 50$ hours in steps of 5 hours. In comparison, plot $\exp(-0.01t)$. Note the slight discrepancy between the two curves.

Calculate the number surviving at $t = 5$ hours by calculating in steps of 1 hour between $t = 0$ and $t = 5$ hours. Why is the discrepancy between this and the plot of $\exp(-0.01t)$ now less than it was before?

3.3. The half-life for the decay of a piece of radio-active material is the time (starting from an arbitrary zero) taken for half of the material to decay. If this time is given as t_1 hours, calculate the equivalent constant failure rate.

3.4. The failure density function of a certain component is as shown in Figure 3.15.

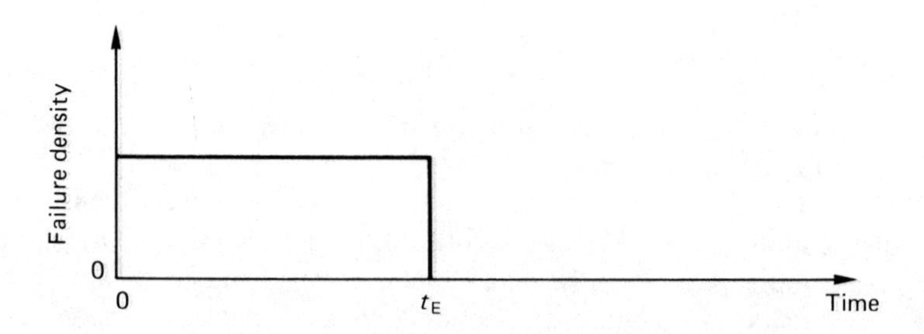

Figure 3.15 The failure density of a component

Calculate $R(t)$ and $\lambda(t)$ for the component and sketch the two curves. What is the MTTF of this component?

4 The Statistics of Failure – II

4.1 Introduction

In this chapter it is shown how failure data can be analysed to find whether or not a component can be considered to have a constant failure rate (CFR). Most electronic components are assumed to have a CFR and thus this is the most important case to consider. Methods of calculating point estimates and confidence intervals for CFR components make up most of the remainder of the chapter. Sections 4.2 to 4.7 inclusive contain general plotting methods. Section 4.8 deals with CFR components only. Section 4.9 indicates briefly how reliability parameters for non-CFR components may be estimated.

4.2 The Weibull Distribution: Mathematical Form

When components have failure rates which cannot be treated as even approximately constant, the mathematical treatment of reliability becomes much more difficult. It has been found that the Weibull distribution is very useful in this context, since it can be made to fit many different curves (including the constant failure rate case).

The three-parameter Weibull reliability function is expressed by

$$R(t) = \exp\{-[(t - \gamma/\eta]^{\beta}\}$$

and the corresponding density function may be found by differentiation to be

$$f(t) = \beta[(t - \gamma/\eta]^{\beta-1}[1/\eta] \exp\{-[(t-\gamma)/\eta]^{\beta}\}$$

It will generally be assumed that $\gamma = 0$ (see below) so that:

$$f(t) = \beta[t/\eta]^{\beta-1}[1/\eta] \exp\{-(t/\eta)^{\beta}\}$$

When $\beta = 1$, $f(t)$ reduces to the exponential failure density function since also putting $\lambda = 1/\eta$ causes the expression to reduce to

$$f(t) = \lambda \exp(-\lambda t)$$

When $\beta < 1$ the Weibull density function becomes infinite at $t = 0$ but when $\beta > 1$ the function is zero at $t = 0$ and has a maximum value at some value of t. Typical curves are shown in Figure 4.1.

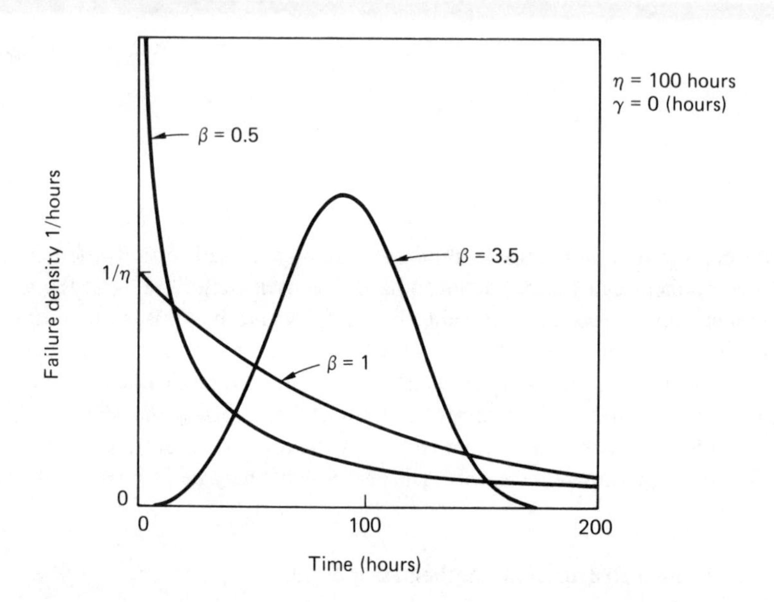

Figure 4.1 Weibull failure density functions showing the effect of varying β

The effect of changing η is merely to cause a change in the scale of t. This effect is shown in Figure 4.2 in which curves of $\beta = 3.5$, $\eta = 100$, 120 and 150 hours are shown. It should be noted that while β is dimensionless, η and γ have the dimensions of time.

The failure rate derived from the Weibull distribution is

$$\lambda(t) = \frac{\beta}{\eta} \; [(t - \gamma)/\eta]^{\beta - 1}$$

It can be seen from this expression that when $\beta < 1$ the failure rate decreases with time; when $\beta = 1$ the failure rate is constant and equal to $1/\eta$; when $\beta > 1$ the failure rate increases with time. Typical curves of failure rate against time are shown in Figure 4.3.

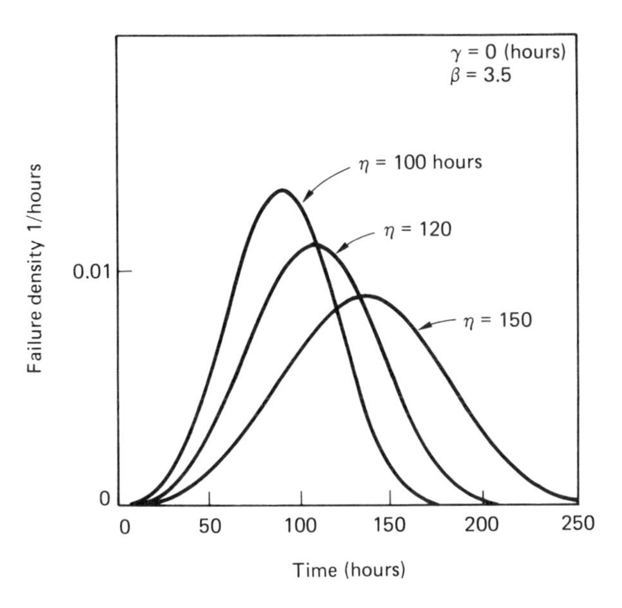

Figure 4.2 Weibull failure density functions showing the effect of varying η

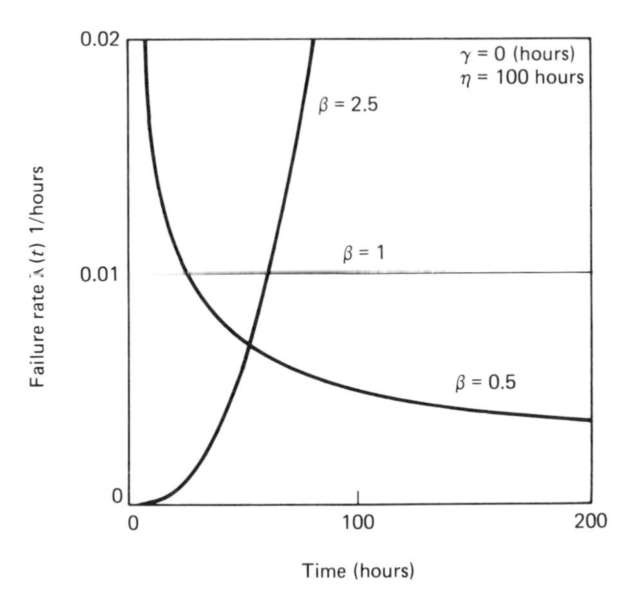

Figure 4.3 Weibull failure rates showing the effect of varying β

4.3 Weibull Plotting: Introduction

In plotting statistical data on failures it is usual to plot failure times on a cumulative graph rather than to try to fit the data to a probability density curve. The reason for this is that by suitable manipulation of the horizontal and vertical scales it is possible to convert an S-shaped cumulative curve into a straight line; it is then much easier to judge whether the experimental data follows a straight line than it would be to judge whether it follows the curve of a probability density function. It is usual to follow this procedure with data which is expected to follow a Normal distribution (using Normal probability paper), and a similar procedure is used for plotting data which may follow a Weibull distribution by using special Weibull probability paper.

4.4 Plotting Data on a Cumulative Graph

If we put, say, five components on test and measure their times to failure then we can list the five failure times in increasing order and plot them on a cumulative graph.

It would be tempting to plot the points as representing the 20 per cent, 40 per cent, 60 per cent, 80 per cent and 100 per cent points of the cumulative distribution. But we must remember that we are trying to show the cumulative failure distribution of the whole statistical population rather than that of a small sample. By plotting the data in the above manner we should imply that 20 per cent of the population times-to-failure will occur before the first data point (t_1) while none of the population times-to-failure will occur after the last data point (t_5). This is clearly an unreasonable supposition. We can use a similar argument to reject using the plotting points, 0 per cent, 20 per cent, 40 per cent, 60 per cent and 80 per cent. The two alternative plotting schemes are shown in Figure 4.4.

For the first scheme the ith point in a total of n data points is plotted as at the $100(i/n)$ per cent point; in the second scheme it is plotted as the $100([i-1]/n)$ per cent point. It can be argued that the best plotting point should lie between these two extremes and so in the case of five data points we might choose the 10 per cent, 30 per cent 50 per cent, 70 per cent and 90 per cent cumulative percentages to correspond to the successive times-to-failure t_1, t_2, t_3, t_4 and t_5. This is not an unreasonable procedure; it corresponds to plotting the data at the $100([i-0.5]/n)$ percentage points. The problem of optimum plotting points is quite complex and for a fuller treatment of this issue the reader is referred to textbooks on statistics (see, for example, Johnson (1964)). For our purposes, we shall merely note that for the ith plotting point (out of n) there is a 'rank' distribution, the mean of which is $i/(n+1)$ (that is, $100i/[n+1]$ per cent) which is sometimes used for plotting. In the plotting of failure data, the median rank is that most widely used. Tables are available giving these and other percentage ranks which are useful in setting up confidence intervals as described below. A very good approximation for the ith median rank from n data points is $100(i - 0.3)/(n + 0.4)$ per cent and this has been used in the construction of Table 4.1 in which mean and median ranks are quoted for the five data points discussed above. A table of median (50 per cent) ranks is included in Appendix 2.

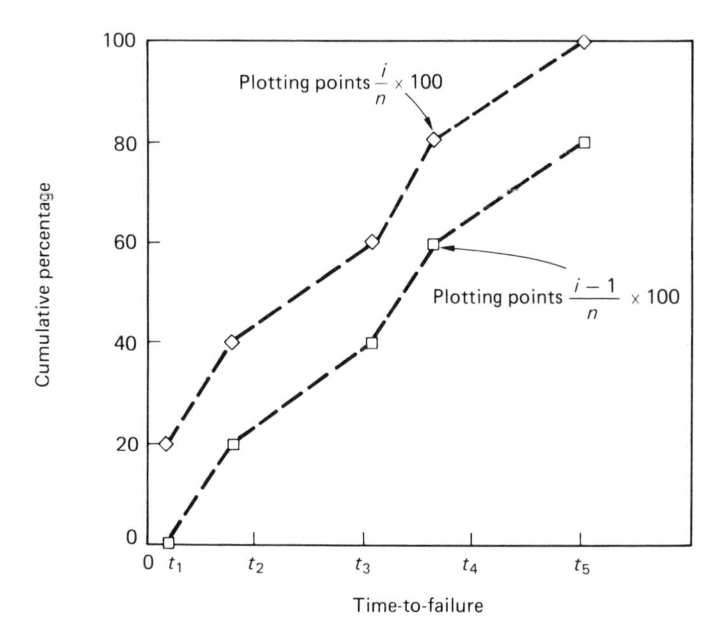

Figure 4.4 Extreme plotting points for five times to failure

Table 4.1 Plotting data for five times to failure

Rank order no. (i)	*Failure time (hours)*	*Mean rank* $\frac{100i}{n+1}$	*Median rank (approx.)*
1	103	16.67	12.96
2	411	33.33	31.48
3	1049	50.00	50.00
4	1327	66.67	68.52
5	2016	83.33	87.04

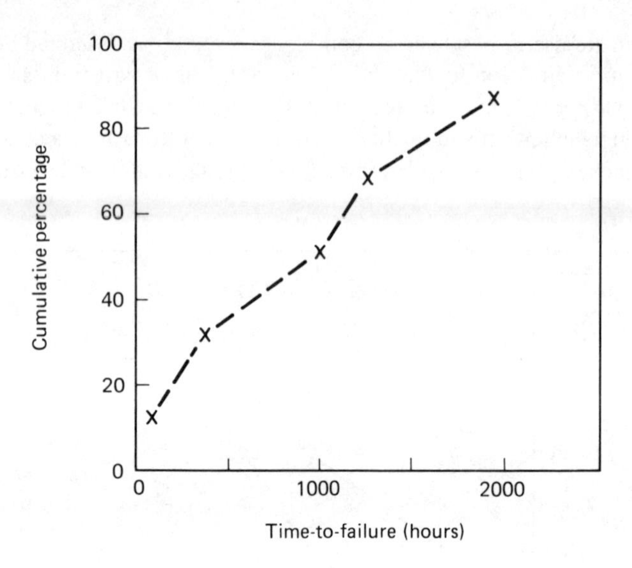

Figure 4.5 Plot of failure data using linear axes

4.5 Weibull Probability Paper

For the Weibull distribution we have

$$F(t) = 1 - R(t) = 1 - \exp\{-[(t - \gamma)/\eta]^{\beta}\}$$

$$(\text{given that } t - \gamma > 0,\ \beta > 0,\ \eta > 0)$$

Re-arranging and taking natural logarithms gives

$$\ln(1 - F(t)) = -[(t - \gamma)/\eta]^{\beta}$$

so that

$$\ln\{1/(1 - F(t)\} = [(t - \gamma)/\eta]^{\beta}$$

and

$$\ln[\ln\{1/(1 \quad F(t))\}] = \beta\ln(t - \gamma) - \beta\ln(\eta)$$

Thus a plot of $\ln[\ln\{1/(F(t))\}]$ against $\ln(t - \gamma)$ is linear and the slope of the line is β. Weibull Probability Paper has a vertical axis which is proportional to $\ln[\ln\{1/(1 - F(t))\}]$ and the horizontal axis is proportional to $\ln(\text{time})$, so that if the data fit a Weibull distribution, the plot on Weibull Probability Paper is a straight line.

The parameter η may be found by noting the fact that when $t - \gamma = \eta$ then $F(t) = 1 - \exp(-1)$ (about 63 per cent) irrespective of the value of β. Thus so long as the plot is linear, the value of $\hat{\eta}$ can be found from the intercept of the plot with the 63 per cent ordinate; this value of the ordinate is usually marked with a dashed line on Weibull Probability Paper.

The parameter γ is called the shift parameter since it represents a simple time shift. In plotting failure data it is most usual to assume that $\gamma = 0$ since a positive value implies that no failure can possibly occur between $t = 0$ and $t = \gamma$ (an assumption which cannot usually be made), while a negative value of γ implies that some failures have occurred before $t = 0$. (At $t = 0$ it follows that $R(t) = \exp[\{-(|\gamma|)/\eta\}^{\beta}]$ whence $R(t) < 1$.) The effect of a non-zero value of γ is to cause the Weibull plot to be curved upwards or downwards at low values of t. The value of γ may be approximated by an iterative process, successively choosing trial values so as to straighten the plot until an optimum value has been obtained. More detailed information on parameter fitting is given in Mann *et al.* (1974).

A computer-generated example of Weibull graph paper is shown in Figure 4.6. The five data points from the previously discussed life-test data are shown plotted on the paper using median ranking. In this figure a straight-line has been fitted to the data points by eye. The value of η is estimated to be 1020 hours.

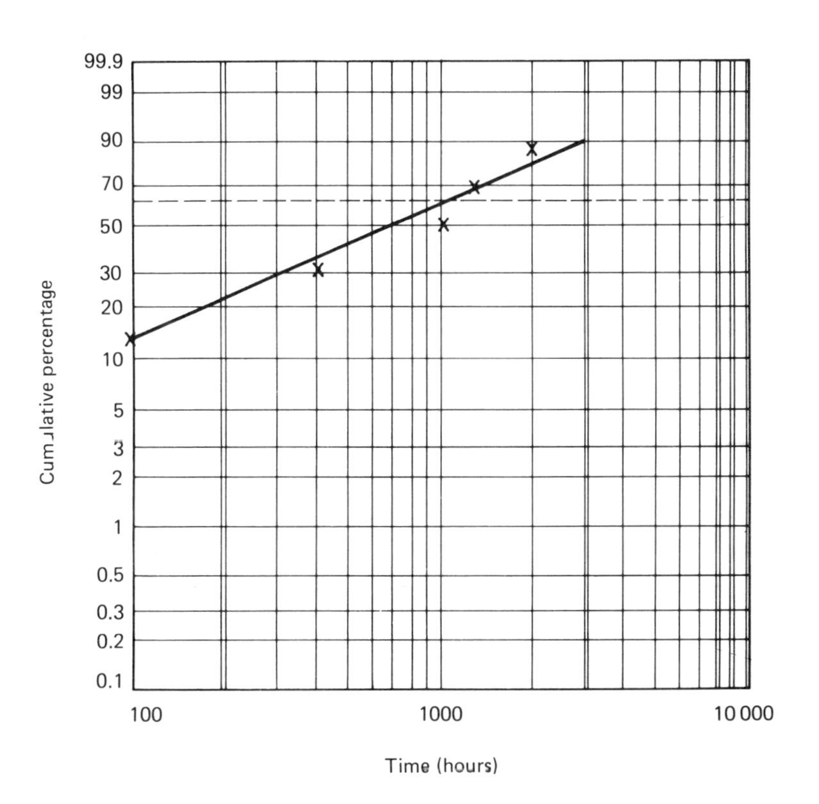

Figure 4.6 A Weibull plot of failure data

The estimated value of β is found by observing that the straight-line fit passes through the points (100 h, 0.13) and (3000 h, 0.90) so that the estimated β (denoted by $\hat{\beta}$) is

$$\hat{\beta} = \frac{\ln[\ln(1/(1-0.90))] - \ln[\ln(1/(1-0.13))]}{\ln[3000] - \ln[100]}$$

$$\approx 0.82$$

Commercially-available Weibull Probability Paper is represented by Chartwell Graph Paper Reference 6572 (Weibull Probability × Log 2 Cycles). The 63 per cent estimation ordinate for η is marked in this paper as in Figure 4.6. The estimated value of β can be found by dropping a perpendicular line from an estimation point on to the straight-line plot. The perpendicular crosses two nomographs: one gives the estimated value of β and the other the cumulative percentage at which the mean value occurs (thus allowing the mean value to be read off the plot). See Figure 4.7.

4.6 Plotting Censored Data

Data which are incomplete in some way are called censored. As an example if 20 components are put on test then it may happen that the test must be ended before all the components have failed. Say that five times to failure are known but all that is known about the other components is that their times to failure are greater than the time at which the test ended. A Weibull plot of this data can still be made using the five known failure times, but these must be plotted using the median ranks for a sample size of 20 ($n = 20$ in the formula of section 4.4) rather than five.

A rather more difficult problem occurs when one or more components are taken off test before they have failed and in between the known times to failure of other components. These data are known as suspensions. It is tempting to ignore the suspensions completely, but to do so would not be making best use of the available data. (Data on component failure are often difficult to acquire!) It is possible to make an allowance for such suspensions by adjustment of the plotting points and the method (see Johnson (1964)) is best illustrated by an example

Consider that 12 components are put on test and the following data obtained:

Times to failure: 1800, 850, 3600, 180, 2400, 460, 6500, 1100 hours

Suspensions occurred at: 1700+, 500+, 2600+, 1350+ hours, where the + sign indicates a suspension instead of a failure

These data are pooled and ordered from lowest to highest to give:

180, 460, 500+, 850, 1100, 1350+, 1700+, 1800, 2400, 2600+, 3600, 6500 hours

as shown in Table 4.2.

Table 4.2 Plotting points for data which include suspensions

Order no.	*Time at failure or suspension (hours)*	*Rank order increment*	*Rank order no. (i)*	*Plotting point (per cent) (median rank; interpolated where necessary)*
1	180	1.00	1	5.6
2	460	1.00	2	13.7
3	500+	1.10	Not plotted	
4	850	1.10	3.10	22.6
5	1100	1.10	4.20	31.5
6	1350+	1.26*	Not plotted	
7	1700+	1.47	Not plotted	
8	1800	1.47	5.67	43.3
9	2400	1.47	7.14	55.2
10	2600+	1.95	Not plotted	
11	3600	1.95	9.09	70.9
12	6500		11.04	86.6

*This increment is not used because it is immediately superseded by that introduced by the next suspension.

The first two data points are plotted as normally, using the first two median ranks from a sample of 12, namely 5.6 per cent and 13.7 per cent. The next time is that of a suspension; this cannot be plotted, but an allowance is made for the fact that this component could have failed at any subsequent time by calculating a new rank increment:

$$\text{New Rank Increment} = \frac{(n + 1) - \text{Previous (plotted) rank order no.}}{1 + (\text{No. of components following current suspension})}$$

where n = total sample size including suspensions

The suspension at 500 hours (500+) thus generates a new rank increment of $(13 - 2)/(1 + 9) = 1.10$. This new increment is used to generate rank order nos of $2 + 1.10 = 3.10$ (at 850 hours) and $3.10 + 1.10 = 4.20$ (at 1100 hours). A suspension occurs at 1350 hours and generates a new increment of (13 – 4.20)/(1 + 6) = 1.26. However this increment is not used because another suspension occurs immediately afterwards at 1700 hours; this causes a new rank increment to be generated of (13 – 4.20)/(1 + 5) = 1.47. The rank order nos for the next two failures (at 1800 and 2400 hours) are thus 4.20 + 1.47 = 5.67 and 5.67 + 1.47 = 7.14 respectively. The suspension at 2600 hours generates a new increment of (13 – 7.14)/(1 + 2) = 1.95 so that the last two failure times (3600 and 6500 hours) have rank order nos of 9.09 and 11.04 respectively (see Table 4.1).

The plotting points in Table 4.2 for non-integral rank order numbers are obtained by linear interpolation in median rank tables or by using the formula $100(i - 0.3)/(n + 0.4)$ as used above but now using non-integral values of i (the rank order number).

The data of Table 4.2 are plotted on proprietary graph paper in Figure 4.7. (90 per cent confidence intervals are also indicated - see section 4.9.) Since the plot shows no curvature at the low end, it is reasonable to take $\hat{\gamma} = 0$ (hours). The value of $\hat{\beta}$ may be found in two ways.

(a) By noting that the plot passes through the two end points, we may calculate it as

$$\hat{\beta} = [\ln[\ln(1/(1 - 0.866))] - \ln[\ln(1/(1 - 0.056))]]/[\ln 6500 - \ln 180]$$
$$= 1.0$$

(b) $\hat{\beta}$ may be obtained from the nomograph in Figure 4.7 by dropping a perpendicular from the reference point on to the straight-line plot. The point at which the perpendicular crosses the nomograph gives the required estimate. (This procedure is only valid so long as the plot can be considered to be straight.) From Figure 4.7 we again obtain $\hat{\beta} = 1.0$.

$\hat{\eta}$ can be estimated by the value of the abscissa at which the plot cuts the 1 – 1/exp (about 63 per cent) ordinate; the required ordinate is indicated in the graph paper by the dashed line labelled 'η Estimator'. From Figure 4.7 we obtain: $\hat{\eta} = 3100$ hours. We thus conclude from the data that the component has a CFR (since $\hat{\beta} = 1$) which is $\hat{\lambda} = 1/\hat{\eta} = 1/3100$/hour.

4.7 Practical Aspects of Data Plotting

It is often desired that a component should be known to have a constant failure rate. In attempting to demonstrate this, it is usual to plot the data using methods described above. However it is very unlikely that practical data will produce a straight-line Weibull plot with a slope exactly equal to unity. Some engineering judgement is needed in deciding whether or not the data adequately demonstrate that the failure rate is constant. The judgement must take into account the amount of data available and what spread could be attributed to the slope of the plot. In this context it is useful to add confidence limits to the plot; see sections 4.8.2 and 4.9.

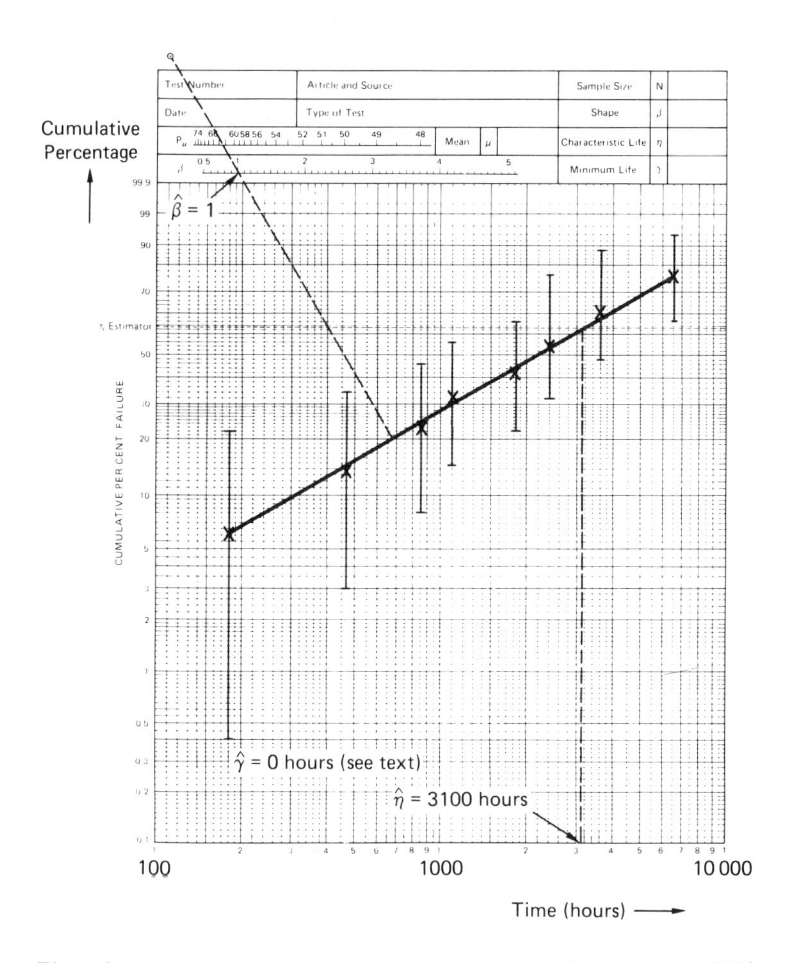

Figure 4.7 A Weibull plot of the data of Table 4.2 using Chartwell Paper 6572

4.8 Point and Interval Estimation for CFR Components

4.8.1 Point Estimates of λ and θ from Failure Data

In general, if a number of components are put on test, there may not be sufficient time available for all the components to fail. As described above, the failure times then form a set of censored data. Tests terminated before all the components fail may be failure-truncated or time-truncated.

Such incomplete data can, of course, only give an estimate of the true failure performance of a component. If there are sufficient grounds for believing that the failure rate is constant, then an estimate of the true failure rate can be made from the available data. Such an estimate is called a point estimate and is designated by $\hat{\lambda}$ to distinguish it from the true, but unknown, failure rate λ.

Similarly, a point estimate of the MTTF is designated by $\hat{\theta}$.

If a large number of components is put on test and the test terminated (either by time- or failure-truncation) before they all fail, then the data obtained may be used to construct a graph such as that shown in Figure 4.8.

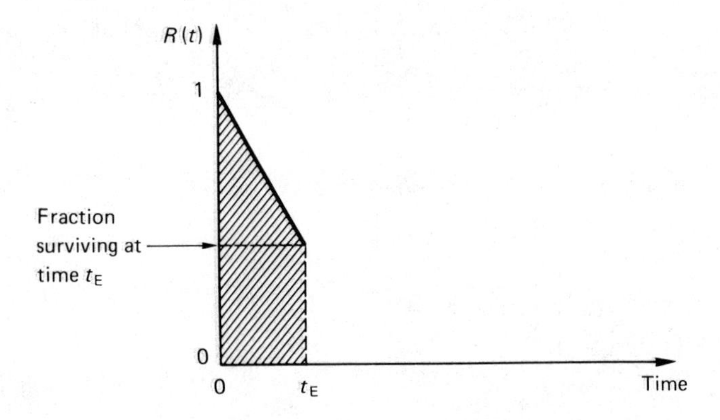

Figure 4.8 $R(t)$ for a truncated test

Two point estimates of λ can now be made:

(a) By noting that $R(t)$ represents the fraction surviving at the time t, we can find after some re-arrangement:

$$\hat{\lambda} = \frac{\ln(\text{fraction surviving at time } t_E)}{t_E}$$

This is not a very accurate estimate of λ since it takes no account of the individual failure times $t_1, t_2 \ldots$

(b) A more accurate estimate (the maximum likelihood estimate) is obtained by noting:

From $t = 0$ to $t = t_E$, area under graph $= \int_0^{t_E} R(t)\,dt$

$$= \int_0^{t_E} \exp(-\lambda t)\,dt = \frac{1 - \exp(-\lambda t_E)}{\lambda}$$

$$= \frac{\text{fraction of components failing from } t = 0 \text{ to } t = t_E}{\lambda}$$

By 'de-normalising' this expression, re-arranging and putting $\hat{\lambda}$ for λ, we obtain:

$$\hat{\lambda} = \frac{\text{number of components failing from } t = 0 \text{ to } t = t_E}{\text{total number of component hours generated from } t = 0 \text{ to } t = t_E}$$

Inversion of this expression gives:

$$\text{estimated MTTF} = \hat{\theta} = \frac{\text{total number of component hours generated}}{\text{number of components failing}}$$

Although these expressions have been obtained by assuming that a known fraction of the total number of components has failed, it is possible to show that these are maximum likelihood estimates from the information:

T = Total component hours generated (by all components whether failed or not)

k = Number of failures

So

$$\hat{\lambda} = \frac{k}{T} \quad \text{and} \quad \hat{\theta} = \frac{T}{k}$$

The similarity between these equations and

$$\text{Expected no. of failures (in time } T) = \lambda T = T/\theta$$

should be noted.

4.8.2 Confidence Intervals

A disadvantage of point estimates is that they give no indication of the amount of data that has been used in making the estimate, and there is thus no direct way of indicating how much reliance one can put on it. As an example, the same point estimate for the MTTF would be obtained from a test having $T = 1000$ hours and $k = 1$ as from a test having $T = 50\,000$ hours and $k = 50$; one would, however, have considerably more faith in the results of the second test than in those of the first. (It is assumed than in both cases a 'good' sample was tested.)

To overcome this limitation the concept of a confidence interval has been developed, and this is often quoted along with the point estimate. A percentage confidence level (100 [1 – α] per cent) is quoted together with the interval, and this has the following meaning: if the same test were repeated a large number of times, then in 100(1 – α) per cent of the tests the calculated confidence interval would embrace the true but unknown value of the parameter (θ or λ in our case). In practice, the test is only made once and the probability that the confidence interval does embrace the 'true' parameter is 0 or 1. When we quote a confidence interval for a test we must hope that the recipient of the information understands that the quoted interval *could* be one of those which does *not* embrace the true value of the parameter.

Confidence intervals can be double- or single-sided. The double-sided interval has at its extreme ends an upper and a lower limit. For the MTTF these limits are designated by θ_{upp} and θ_{low} with corresponding expressions for λ. The single sided interval can have an upper confidence limit (in which case the interval is from zero to θ_{upp}) or a lower confidence limit (in which case the interval is from θ_{low} to infinity).

The meaning of confidence intervals is illustrated below in Figures 4.10 and 4.11. As is to be expected, the more data that are available, the narrower the confidence interval and vice versa. The confidence level (the value of [1 – α] as a percentage) is usually 60 per cent or greater.

The approach to interval estimation outlined above is the 'classical' method developed by Neyman. It is possible to use Bayesian methods of interval estimation (see, for example, Mann *et al.* (1974) or fiducial intervals (see Kendall and Stuart (1973)). Calculation of the confidence interval depends upon whether the test is time-truncated (that is, stopped at a pre-arranged time) or failure-truncated (that is, stopped at the moment when a certain pre-arranged number of failures has occurred).

4.8.3 Double-sided Confidence Intervals for Failure-truncated Tests

It can be shown that if a failure-truncated test is carried out in which the truncation was at the kth failure, then $2k\hat{\theta}/\theta$ is distributed as the statistical probability density function (pdf) χ^2 with $2k$ degrees of freedom. See Mann *et al.* (1974). The general shape of the χ^2 distribution is shown in Figure 4.9.* The exact shape is different for each individual value of the degrees of freedom (n); graphs for $n = 4$ and $n = 10$ are shown in the figure.

*For a definition of the χ^2 function, see Appendix 1.

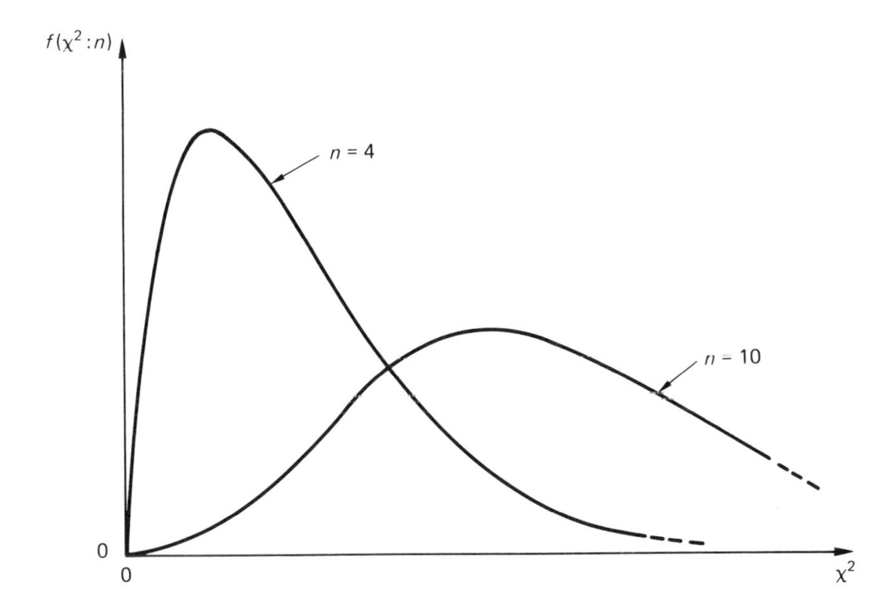

Figure 4.9 The χ^2 distribution shown for two particular numbers of degrees of freedom ($n = 4$ and $n = 10$)

Since $\theta = T/k$, it follows that $2T/\theta$ is distributed as χ^2 with $2k$ degrees of freedom. Since we are dealing here only with CFR components, the total component hours which are generated can be broken down into a manner similar to that shown in Figure 3.10 except that time T in part (a) must coincide with a failure. We do not know the 'true' value of θ; but we can ask the question 'what large value of θ is such that there is only a small ($\alpha/2$) probability that we should have found a value of T as low as or lower than this?'. The answer is obtained by equating the value of χ^2 which cuts off an area of $\alpha/2$ on the left-hand side of Figure 4.10 to the value of $2T/\theta_{\text{upp}}$, where θ_{upp} is the upper confidence limit of θ. Values of χ^2 are quoted in tables (see Appendix 1) in which the area (γ) cut off to the right of the abscissa is given together with the relevant degrees of freedom.

Thus the upper confidence limit θ_{upp} can be obtained by putting

$$\frac{2T}{\theta_{\text{upp}}} = {}_{1-\alpha/2}\chi^2_{2k}$$

which is equivalent to

$$\theta_{\text{upp}} = \frac{2T}{{}_{1-\alpha/2}\chi^2_{2k}}$$

χ^2 is quoted with two subscripts (the positioning and labelling of subscripts in the χ^2 distribution varies from author to author and the reader is advised to check carefully each author's terminology): the right-hand subscript ($2k$) refers to the degrees of freedom of χ^2 and the left-hand subscript (the area parameter) refers to the area $(1 - \alpha/2)$ cut off to the right of the given abscissa - see Figure 4.10. (Since χ^2 is a probability density function, the total area under the curve is unity.)

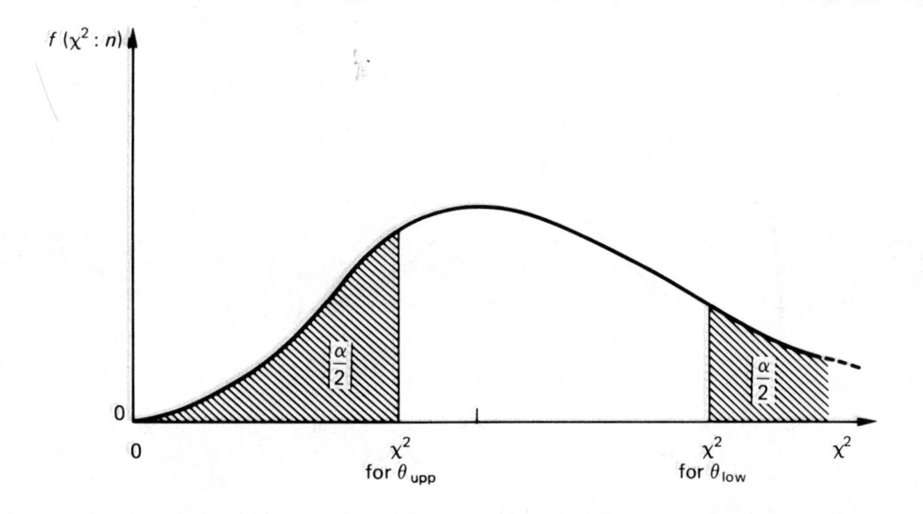

Figure 4.10 The χ^2 distribution showing areas of $\alpha/2$ cut off to the right and to the left

The lower limit θ_{low} is found in a similar fashion by asking the question 'what low value of θ is such that there is only a small ($\alpha/2$) probability that we should have found a value of T as high as or higher than this?'. A similar development to that above leads to

$$\theta_{low} = \frac{2T}{{}_{\alpha/2}\chi^2_{2k}}$$

since the area cut off to the right in Figure 4.10 is now $\alpha/2$.

By reference to the χ^2 tables of Appendix 1 it is found that when $k = 1$ and $T = 1000$ hours, the 90 per cent* $(100[1 - \alpha])$ upper and lower confidence limits are

$$\theta_{upp} = \frac{2000}{0.103} = 19\,400 \text{ hours} \quad \theta_{low} = \frac{2000}{5.99} = 330 \text{ hours}$$

*Solving $90 = 100[1 - \alpha]$ gives $\alpha = 0.1$ and thus $\alpha/2 = 0.05$. For the upper confidence limit, the χ^2 tables are entered with area parameter (γ) of 0.95; for the lower limit it is 0.05.

When $k = 50$ and $T = 50\,000$ hours, the 90 per cent confidence limits are

$$\theta_{upp} = \frac{100\,000}{77.9} = 1280 \text{ hours} \quad \theta_{low} = \frac{100\,000}{124.3} = 800 \text{ hours}$$

The narrowing of the interval as a result of the greater amount of data is readily seen in the above example; the confidence level ($100[1-\alpha]$) in each case is 90 per cent.

The meaning of the double-sided confidence interval is illustrated in Figure 4.11 in which a confidence level of 80 per cent has been assumed, so that in the long run 4 out of 5 of the intervals shown will embrace the true but unknown value of θ.

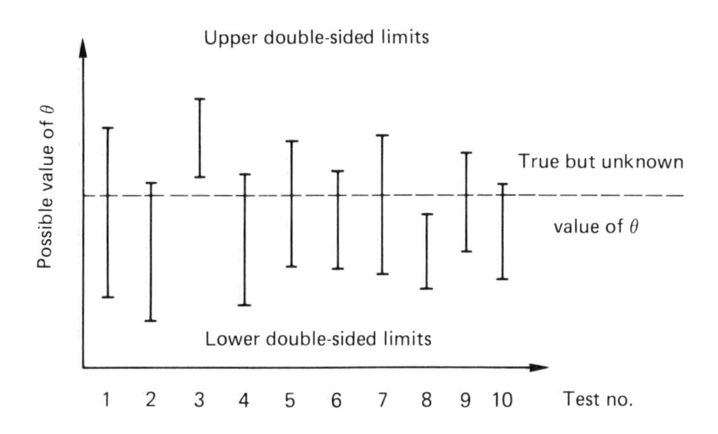

Figure 4.11 Illustration of the meaning of a double-sided confidence interval

Since θ and λ are inversely related, it follows that

$$\lambda_{upp} = \frac{1}{\theta_{low}} = \frac{_{\alpha/2}\chi^2_{2k}}{2T} \quad \text{and } \lambda_{low} = \frac{1}{\theta_{upp}} = \frac{_{1-\alpha/2}\chi^2_{2k}}{2T}$$

4.8.4 Single-sided Confidence Intervals for Failure-truncated Tests

Single-sided upper and lower confidence limits may be obtained by allowing the whole of the area α to lie to the left or right respectively in Figure 4.8. The meaning of single-sided confidence limits on θ is illustrated in Figure 4.12. The confidence intervals for θ are now from 0 to θ_{upp} and from θ_{low} to infinity. As in Figure 4.11 a confidence level of 80 per cent is assumed, so that in the long run 4 out of 5 of the intervals will include the true but unknown value of θ.

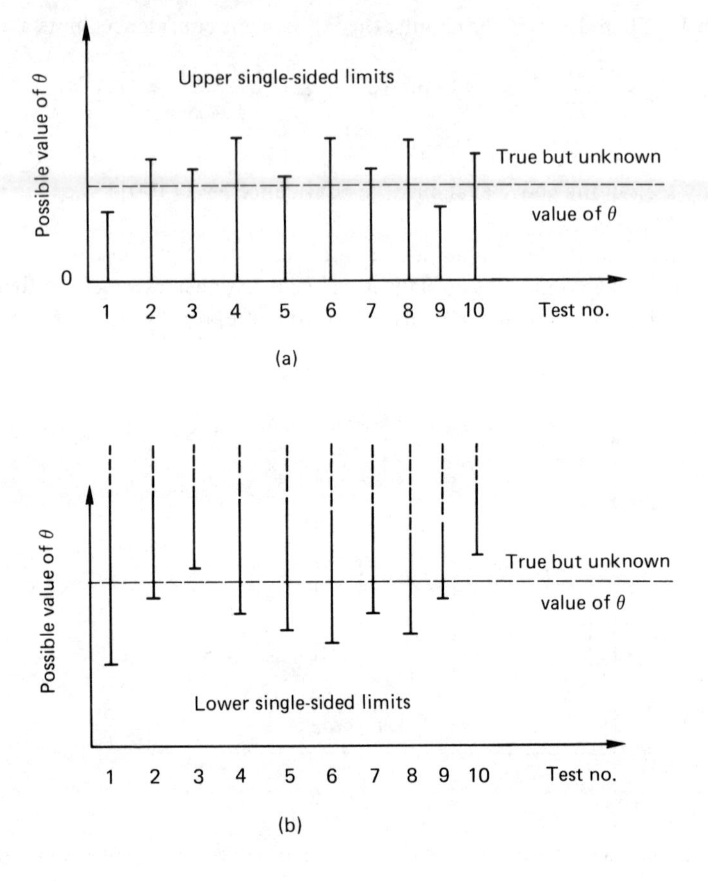

Figure 4.12 Illustrations of the meaning of single-sided confidence intervals: (a) upper single-sided confidence interval; lower single-sided confidence interval

Following the methods of the previous section, the single-sided upper and lower confidence limits of θ are given by

$$\theta_{\text{upp}} = \frac{2T}{{}_{1-\alpha}\chi^2_{2k}} \qquad \theta_{\text{low}} = \frac{2T}{{}_{\alpha}\chi^2_{2k}}$$

The single-sided upper and lower confidence limits for λ are given by

$$\lambda_{\text{upp}} = \frac{1}{\theta_{\text{low}}} = \frac{{}_{\alpha}\chi^2_{2k}}{2T} \qquad \lambda_{\text{low}} = \frac{1}{\theta_{\text{upp}}} = \frac{1 - {}_{\alpha}\chi^2_{2k}}{2T}$$

4.8.5 Double-sided Confidence Intervals for Time-truncated Tests

It can be shown that $2k\hat{\theta}/\theta$ is *approximately* distributed as χ^2 (see Mann *et al.* (1974)). The estimation of θ_{upp} is made as before with $2k$ degrees of freedom but the estimation of θ_{low} is now made with $2(k + 1)$ degrees of freedom. An easy way to remember this modification is to consider that a pessimistic view of the outcome of the test would be that it terminated an instant before a failure would have occurred, and thus for estimating the lower confidence limit an extra failure is added on to the actual number (k). It should be noted that as a result of this modification, θ_{low} can be estimated when no failures have occurred.

Thus

$$\theta_{low} = \frac{2k\hat{\theta}}{_{\alpha/2}\chi^2_{2(k+1)}} = \frac{2T}{_{\alpha/2}\chi^2_{2(k+1)}}$$

$$\theta_{upp} = \frac{2k\hat{\theta}}{_{1-\alpha/2}\chi^2_{2k}} = \frac{2T}{_{1-\alpha/2}\chi^2_{2k}}$$

In the case of failure rates:

$$\lambda_{upp} = \frac{1}{\theta_{low}} = \frac{_{\alpha/2}\chi^2_{2(k+1)}}{2T}$$

$$\lambda_{low} = \frac{1}{\theta_{upp}} = \frac{_{1-\alpha/2}\chi^2_{2k}}{2T}$$

The total component hours T can be generated by one continuous test or by a number of parallel tests as shown in Figure 3.12.

4.8.6 Single-sided Confidence Intervals for Time-truncated Tests

Single-sided confidence limits may be found as in the failure-truncated case. The formulae become

$$\theta_{low} = \frac{2T}{_{\alpha}\chi^2_{2(k+1)}} \qquad \theta_{upp} = \frac{2T}{_{1-\alpha}\chi^2_{2k}}$$

and

$$\lambda_{upp} = \frac{_{\alpha}\chi^2_{2(k+1)}}{2T} \qquad \lambda_{low} = \frac{_{1-\alpha}\chi^2_{2k}}{2T}$$

4.8.7 Summary Table of the Use of the χ^2 Distribution for Calculation of Confidence Limits on θ

Table 4.3 gives a summary of the values of the χ^2 distribution which should be used for the calculation of confidence limits on θ in failure- or time-truncated life tests. It should be used only for calculating confidence limits for CFR components. Values of χ^2 are given in Appendix 1.

Table 4.3 Use of χ^2 distribution in calculating confidence limits on θ for CFR components

		Limit	*Test*	
			Failure-truncated	*Time-truncated*
Confidence interval	Single-sided	Upper	${}_{1-\alpha}\chi^2_{2k}$	${}_{1-\alpha}\chi^2_{2k}$
		Lower	${}_{\alpha}\chi^2_{2k}$	${}_{\alpha}\chi^2_{2(k+1)}$
(Confidence level of $1-\alpha$)	Double-sided	Upper	${}_{1-\alpha/2}\chi^2_{2k}$	${}_{1-\alpha/2}\chi^2_{2k}$
		Lower	${}_{\alpha/2}\chi^2_{2k}$	${}_{\alpha/2}\chi^2_{2(k+1)}$

This table can also be used for calculating confidence intervals on failure rates but it must then be remembered that

$$\lambda_{upp} = 1/\theta_{low} \quad \text{and} \quad \lambda_{low} = 1/\theta_{upp}$$

4.9 Confidence Intervals on Graphical Plots

Confidence intervals may be added to graphical plots; the method is distribution-free and so can be applied to CFR and non-CFR components. The data in Table 4.2 are reproduced in Table 4.4 to which have been added the 5 per cent and 95 per cent ranks (given in Appendix 2) for each data point; interpolation has been used where necessary. Plotted data from Table 4.4 are shown above in Figure 4.7. The 90 per cent confidence interval for each data point is the vertical distance between the 5 per cent and 95 per cent ranks. There are two different methods described in the literature for constructing confidence intervals on plotted data. Reasons for using the method shown here are given in the draft British Standard BS 5760 - Part 2 (1986), paragraph 3.2.5.6. Other confidence intervals can be obtained in this way; for example, 10 per cent and 90 per cent ranks would give an 80 per cent confidence interval.

Table 4.4 Failure data from Table 4.2 with 5 per cent and 95 per cent ranks added. Ranks are given as percentages and are interpolated where necessary

Order no.	*Time at failure or suspension (hours)*	*Rank order no. (i)*	*Median rank*	*5 per cent rank*	*95 per cent rank*
1	180	1	5.6	0.4	22.1
2	460	2	13.7	3.0	33.9
3	500+	–	–	–	–
4	850	3.10	22.6	7.7	44.7
5	1100	4.20	31.5	14.8	54.3
6	1350+	–	–	–	–
7	1700+	–	–	–	–
8	1800	5.67	43.3	22.4	66.0
9	2400	7.14	55.2	32.6	76.4
10	2600+		–	–	–
11	3600	9.09	70.9	48.1	88.3
12	6500	11.04	86.6	66.6	97.1

4.10 Parameter Estimation in the Non-constant Failure Rate Case

4.10.1 Weibull Distribution

If the data can be shown to follow a Weibull distribution, then estimates of the parameters γ, η and β can be found and the point estimate of the MTTF is given by

$$\hat{\theta} = \hat{\gamma} + \hat{\eta}\,\Gamma\,(1 + 1/\hat{\beta})$$

where $\Gamma(\cdot)$ represents the mathematical gamma function: see Abramowitz and Stegun (1964).

Methods of calculating confidence limits on θ are given in Mann *et al.* (1974).

4.10.2 Non-Weibull Distributions

If the data do not fit a Weibull distribution, then a fit to some other distribution (lognormal etc.) should be sought. Methods of point and interval estimation exist for most named distributions. (For the lognormal case see Aitken and Brown (1976).)

4.11 Reliability as a Function of Other Variables

It is sometimes more appropriate to make reliability a function of 'no. of operations', 'cycles' or 'working time' rather than always a function of elapsed time. Doing this may be more realistic than using just elapsed time, but some care is needed over the ensuing predictions since there may then be 'hidden' failure modes which are not taken into account. For example, the reliability of a switch may be taken as a function of the number of operations. In practice, it might happen that corrosion is a significant cause of failures and a reliability prediction based solely upon number of operations could be badly in error, particularly if the switch is infrequently used.

4.12 Trend Analysis

Consider that a failure-truncated test was made using one test bay in which one component at a time were tested, failed components being successively replaced by new components. The test was truncated at the 5th failure when the following times to failure had been measured (in hours): 6553, 3707, 2218, 1206, 442. If the data are re-ordered from lowest to highest and plotted on a Weibull graph they will be found to fit a straight line exactly with a slope of unity. An incautious analyst might conclude that the component had a CFR and measure a value of MTTF from the graph of 3200 hours. However, it will be seen that the successive times to failure are steadily decreasing and this information was destroyed by the re-ordering of the data as was required for the Weibull plot. There is only one chance in 5! (that is, 120) that this particular sequence was obtained by chance, and so the wise analyst would make note of this fact and attempt to find a reason for it (possibly a worsening with time of the production process). It is thus always worthwhile (if possible!) to examine the unordered data for a trend, whether upwards or downwards. Sophisticated statistical methods for trend analysis are available, although a simple bar chart would be sufficient indication of a problem as shown in Figure 4.13.

4.13 Non-central and Randomised Confidence Intervals

The double-sided confidence intervals described in section 4.8 are called central intervals because equal areas have been allocated to both the left-hand and right-hand areas cut off in Figure 4.10. Owing to the non-symmetrical form of the χ^2 distribution, a slightly shorter interval might be obtained (for the same confidence level) by allowing the two areas to be different. A confidence interval obtained in this way is called a non-central interval; these are not normally used in reliability engineering since the slight shortening of the interval so obtained is not worth the effort involved.

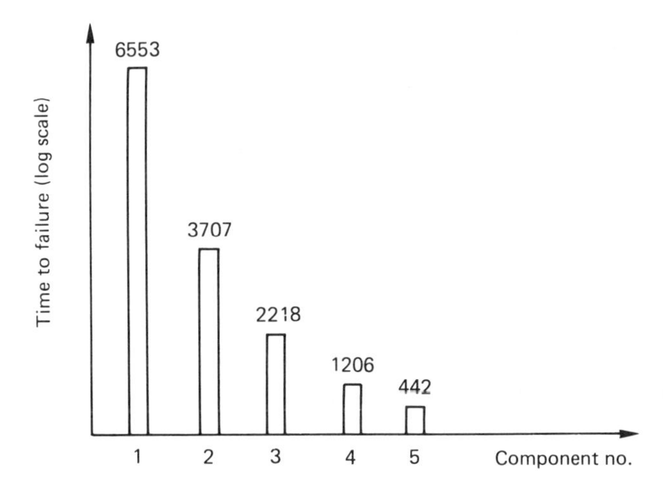

Figure 4.13 Bar chart showing trend in times to failure

It may sometimes be useful to calculate a randomised confidence interval; for details the reader should consult a statistical textbook (such as Kendall and Stuart (1973)).

Exercises

4.1. Find $\hat{\theta}$ and the lower single-sided 90 per cent confidence limit on θ for the failure-truncated life test with the following data:

Number on test = 44
Number of failures (k) = 5
Times of failure = 400 hours
800 hours
1200 hours
1600 hours
1900 hours

You are given that ${}_{0.10}\chi^2_{10} = 16.0$.

It is assumed that the component under test has a constant failure rate.

4.2. Find the upper and lower 90 per cent double-sided confidence limits on λ for the time-truncated life test with the following data:

Number on test = 10
Number of failures (k) = 4

Times of failure	= 230 hours
	500 hours
	940 hours
	1250 hours
Time of truncation	= 1300 hours

Tables of χ^2 are given in Appendix 1. It is assumed that the component under test has a constant failure rate.

4.3. You are offered a choice of two competing devices, both of which have constant failure rates. The first device (A) has a mean time to failure of 10 000 hours. The second device (B) has a reliability of 0.99 over a mission time of 1000 hours. Would you prefer device A or device B?

4.4. Find the lower 60 per cent single-sided confidence limit for the MTTF of a component given the following data:

Number of failures (k) = 0
Number of component hours generated = 100 000

The component is assumed to have a constant failure rate.
Why is it not possible to define $\hat{\theta}$ and the 60 per cent upper confidence limit of θ?

4.5. Choose 10 times to failure at random from an exponential distribution with $\lambda = 1 \times 10^{-4}$/hour. Plot these times on Weibull Probability Paper using median ranks. Fit the best straight line that you can to the data and estimate the values of β and η. Compare these values with those which you would expect them to have.
[The random times to failure (t_i) can be chosen by taking 10 random numbers in the range 0 to 1 (using a calculating machine or random number tables) and then finding the t_i from the equation

$$n_i = 1 - \exp(-\lambda t_i)$$

where n_i is the ith random number.]

4.6. Using the fact that $\Gamma(5) = 4! = 24$, plot the graph of χ^2 with 10 degrees of freedom and mark the abscissas which cut off areas of 0.20 to the right and to the left of the distribution.

4.7. A CFR component is tested with the following results:

No. of failures (k) = 5
No. of component hours generated (T) = 100 000 hours

The test was time-truncated.
Calculate the 90 per cent 80 per cent and 60 per cent single-sided lower confidence limits on the MTTF (θ).

5 Component Reliability

5.1 Introduction

For our purposes we define a component to be 'a non-repairable device'. This definition is somewhat elastic. For example, an altenator in a car may in some circumstances be taken as non-repairable (when it fails, it is removed and thrown away) and hence classed as a component. However, if spares are difficult to obtain, then an attempt may be made to repair it. It then becomes repairable and should be considered a subsystem rather than a component. Nevertheless, for reliability purposes the above definition is adequate since reliability predictions are usually made using well-known sources of data, and 'component' may be taken to mean any device which is listed in one of these sources.

5.2 Variation of Failure Rate with Time

The general form of the failure rate curve with time (for a particular level of stress) is the bathtub curve of Figure 3.7. However, most components (and especially electronic components) are treated as if they had constant failure rates; these rates are quoted in handbooks of failure rate data, see section 5.12. Early and wear-out failures can usually be ignored because:

(a) components and/or equipments are often burnt-in so that few early failures should occur;
(b) reliability predictions are strictly valid only for mature equipments (see Chapter 17 on Reliability Growth);
(c) wear-out of the components usually occurs well beyond the design life of the equipment; alternatively, limited-life* components are replaced well before wear-out occurs.

*A Limited-life component is one which has a Useful Life less than, or not much greater than, the Design Life of the equipment containing it. The Design Life of telecommunications equipment is commonly 15–20 years although it may be less.

These assumptions need rigorous checking in any particular case. For example, in a missile system, infant failures may be catastrophic whereas wear-out failures are unlikely to happen.

For some components it might be more appropriate to give the failure rate as a function of number of operations rather than of time. HRD4 (see section 5.12) quotes the failure rates of relays as functions of time but does also give an upper or lower limit on the number of operations per year.

5.3 Early and Wear-out Failure Mechanisms

5.3.1 Early Failure Mechanisms

Early failure is generally caused by (a) poorly-executed manufacturing processes (such as badly-welded joints in resistors; badly-formed internal bonds in ICs); or (b) faulty materials (such as pinholes in the insulating plastic sheet in capacitors).

Although some authorities claim that early failures can occur during the first 2 years of usage, it seems to be generally accepted that early failures will usually be confined to the first 6 months (or even 2 months).

5.3.2 Wear-out Failure Mechanisms

In mechanical and electro-mechanical components, wear-out is most obviously caused by friction between moving parts (such as wear in bearings or the removal of gold plating on the conducting surfaces of relays); fatigue (in both metals and plastics); and corrosion.

The following is a selection of components and their wear-out mechanisms (see also Amerasekera and Campbell (1987)):

ICs	Ion migration
†Lasers	Gradual degredation
Resistors	(a) Electrical bridging in spirally-cut resistors (b) Evaporation of volatile chemical constituents leaving the resistor brittle and liable to crack
†Electron tubes	Loss of cathode emission
†Secondary batteries	Loss of capacity
†Wet Al electrolytic capacitors	Drying-out caused by diffusion of the electrolyte

Components marked † are generally regarded as having a limited life.

5.4 Other Failure Mechanisms

5.4.1 General

As described in section 3.3, most failures during the constant failure rate period of the bathtub curve occur as the result of a random accumulation of internal stresses, an instantaneous peak of which causes the component to fail, see Figure 3.1. Such random peaks will always occur and the only way to reduce the probability of failure is by making the component stronger or by improving its quality. (Improving the quality of a component presumably decreases the root-mean-square of the stress waveform of Figure 3.1. It is understood that the more reliable a component is, the less noisy it is likely to be; there thus appears to be an interesting correlation between low internal noise and low variation in internal stresses.)

A 'soft' failure in a semi-conductor memory is a change in a stored bit caused, it is believed, by an alpha-particle. This is not a catastrophic failure because the memory is still usable after the error has occurred. The failure rate for soft errors is believed to be low; their effects can be minimised by the use of error-detecting and error-correcting digital codes (see section 16.4.10).

5.4.2 Failure during Storage

Most components are unlikely to fail during storage (that is, while no power is applied and they can be considered to be unstressed). However some components may suffer from corrosion if they are not well protected, and atmospheric attack of this sort can cause failure during storage. Some types of CMOS Integrated Circuit are believed to be prone to failure during storage owing to the ingress of moisture up the leads and so into the package. Another possible cause of failure is fungus attack in humid tropical conditions. Equipment which will be used in such an environment should be tropicalised.

In any situation where equipment is subjected to long storage periods or to exceptional ambient conditions (high or low temperatures, vibration etc.) during storage, an attempt should be made to assess the probability of failure during this period. It is worthwhile remembering that the *Challenger* shuttle disaster was caused by an O-ring being subjected (before launch) to a lower temperature than that for which it had been designed.

5.5 Techniques of Burn-in and Early Replacement

5.5.1 Burn-in

Many component manufacturers subject their components to burn-in before releasing them to their customers. The components are stressed for a given period of time during which, it is hoped, the majority of infant failures will appear. The surviving components are then shipped to the customer, and these components should then have a constant failure rate equal to that at the bottom of the bathtub curve. In order to save time, components are usually stressed during burn-in at higher levels than those which they would normally experience.

In practice, burn-in is a high extra expense for the manufacturer who consequently makes the burn-in period as short as he can. Burn-in always adds to the price of components, and the customer must decide whether the extra cost of the components is worthwhile.

Further information on burn-in can be found in Jensen and Peterson (1982).

5.5.2 Early Replacement

A policy of early replacement is one which replaces limited-life components before their failure rates begin to rise sharply at the onset of wear-out. There is an obvious cost to be born in the maintenance work involved and whether or not this is deemed worthwhile depends upon how critical a failure might be, both from the cost and safety aspects.

If the failure density of wear-out can be modelled by a normal (Gaussian) distribution, then it may be decided that components should be replaced at a time between $m - 2s$ and $m - 3s$, where m = mean time to wear-out and s = standard deviation of time to wear-out.

5.6 Factors which Modify Component Failure Rates

5.6.1 Introduction

Component failure rates are dependent upon a number of factors, chief of which are temperature and stress. The temperature which a component experiences is dictated both by the ambient temperature and the power generated within the component itself. Arrhenius's equation is widely used for the extrapolation of failure rates from one temperature to another. Component failure rates are also affected by other environmental factors apart from temperature; these are chiefly vibration (including acceleration), humidity and dust. In general, the more stressful the environment the higher the failure rate. The environment may vary during the design life of the system containing the component; this variation is shown in the system's mission profile (see Section 2.2) and the component's failure rate will change as the environment changes. Other factors which affect failure rates include the component quality, maturity of the manufacturing process, device complexity etc.

5.6.2 *Temperature*

The failure rate of a component is highly temperature-dependent and it is usually assumed for constant failure rate components that the dependency follows Arrhenius's equation:

$$\lambda = K[\exp(-E/kT)]$$

where

λ = component failure rate
K = a constant
E = activation energy of failure process (usually given in electron-volts)
k = Boltzmann's constant = 8.63×10^{-5} eV/K

T = temperature (K).

The equation is particularly useful for deriving a failure rate at one temperature (say T_2 K) when the failure rate at another temperature (say T_1 K) is known; since

$$\frac{\lambda(T_2)}{\lambda(T_1)} = \exp\left[\frac{E}{k}\left(\frac{1}{T_1} - \frac{1}{T_2}\right)\right]$$

$\lambda(T_2)$ can be obtained from this equation so long as the activation energy (E) for the failure process is known.

Activation energies for packaging failures lie in the range 0.25–0.5 eV, while those for semi-conductor chip failures lie in the range 0.5–1.5 eV. Thus, if Arrhenius's equation is to be used, some compromise activation energy must be used, since failures of a component normally consist of a number of different modes. In telecommunication and computer systems, an activation energy of 0.5 eV is often assumed as a compromise value for calculating the temperature dependence of component or board failure rates.

The value of the absolute temperature used in Arrhenius's equation should be the highest temperature inside the component (for example, the junction temperature of a bipolar transistor) rather than the ambient temperature of the air around the component, since it is that temperature which influences the failure rate. As an example, the ambient temperature inside a telephone exchange may be 25°C; if air at this temperature is blown upwards through a rack of printed circuit boards, there may be a temperature rise of 10 °C from top to bottom of the board. A bipolar transistor mounted at the centre of the board will be surrounded by air at approximately 30 °C (that is, mid-way between 25 °C and 35 °C). The junction of the transistor may be 20 °C above the external case temperature, so that the temperature to be used in the Arrhenius equation would in this case be 50 °C (323 K).

Table 5.1 lists the factors by which the failure rate at 30 °C should be multiplied to give the failure rate at higher temperatures; the calculation is based upon an activation energy of 0.5 eV.

Table 5.1 Acceleration factors based upon 0.5 eV activation energy and 30 °C

Temperature (°C)	30	40	50	60	70	80
Acceleration factor	1	1.8	3.3	5.6	9.3	15.0

5.6.3 Environment

Environmental factors which may affect a component's failure rate include:

ambient temperature
humidity
salt (causes corrosion, particularly when allied with high humidity)
vibration
acceleration (mechanical shock)
radiation
low pressure (experienced in space vehicles)
dust.

Since there are infinitely many possible variations of environment, it is usual in reliability engineering to select the most appropriate from a number of categories such as Ground, Benign; Ground, Fixed; Space, Flight etc. (see section 5.12.2).

5.6.4 Quality Level and Cost

The Quality Level* of a component is characterised by the combination of tests, screening and release conditions for that particular Quality Level.

The effect of Quality Level on the component failure rate is reflected in a multiplying factor π_Q (see section 5.12.2). HRD4 (see section 5.12) defines three Quality Levels (3 being the highest, 1 the lowest) according to BT definitions. The higher the Quality Level the more reliable is the component; for Silicon Integrated Circuits the π_Q multiplying factor in HRD4 is 0.5, 1.0 and 2 for Quality Levels 3, 2 and 1 respectively.

Quality Levels in MIL-HDBK-217E are denoted by letters. For example Microelectronic Devices have the following Quality Levels: S, S-1, B, B-1, B-2,

*It should be noted that the term 'Quality Level' is defined differently in Quality Assurance usage, in which it means what is called 'percentage defective' in Figure 18.1.

D and D-1. The corresponding values of π_Q range from 0.25 (for S) to 20.0 (for D-1). Discrete Semiconductors have Quality Levels JANTXV, JANTX, JAN and Lower; the range in π_Q is from 1.0 (for JANTXV) to 5.0 (for Lower).

The higher the Quality Level, the more expensive is the component. As a rough guide to component cost, a passive component for an undersea system might cost in the region of £50; a similar component for a mass-produced system would cost (when purchased in bulk) less than 1p. A similar comparison for active components shows a range of about £1000 to a few tens of pence.

In general, the reliability of a component is reflected in its cost; although it should be noted that the makers of ICs for digital watches manage to produce a very reliable component at a remarkably low cost. Quality Levels in the UK are also defined in BS 9000. It is sometimes difficult to relate the Quality Level definitions in the different documents. Some information is given in the MoD document DEF-CON 17.

5.6.5 Stress

It is usual to under-stress a component in order to ensure that its failure rate is not excessively high. For power-limited devices (such as transistors) the designer is usually limited to not more than 50 per cent of the maximum quoted power. For voltage- or current-limited devices (such as capacitors) the limit is usually 70 per cent of the maximum. Computer programs used for reliability predictions (see sections 5.13 and 19.4) contain default conditions relating to the stress on components. In failure rate databases (see section 5.12) some formulae contain power or voltage derating factors.

5.6.6 Other Factors

Other factors which affect component failure rates and which are included in the formulae for the calculation of failure rates include Complexity and Maturity of Manufacturing Process. The latter factor is based upon historic data which indicate that component reliability improves as a manufacturing process matures. Presumably in the early stages of development of a new manufacturing process there are blemishes which affect component reliability. As the process matures the blemishes are gradually removed and the component failure rate decreases.

5.7 Techniques for Improving Component Reliability

Apart from the use of burn-in and early replacement (where these techniques are appropriate) component reliability may be improved by:

Improving the ambient conditions by	cooling* (use of fans or air-conditioning) air-conditioning (temperature control, humidity control) air filtration (removal of dust by filters) anti-vibration mountings
Protecting the component by	anti-corrosion treatment anti-fungus treatment radiation hardening

The design of reliable components is discussed below (see section 5.9).

5.8 Component Failure Modes

5.8.1 Separation of Modes

Hitherto, no distinction has been drawn between different component failure modes. The failure rate has been defined in terms of catastrophic failures. But it must be recognised that components may have multiple failure modes, such as failure into short-circuit, failure into open-circuit, excessive parameter drift etc.

For simplicity, in what follows only two failure modes are considered; the treatment can easily be extended to more failure modes if required.

It has been shown that if the overall failure rate is λ (a constant) then in T component hours λT failures can be expected. When considering just two failure modes, it seems reasonable to expect that λ can be broken down into two constant failure rates λ_o (open-circuit) and λ_s (short-circuit) which are themselves both constant.

Thus, with λ_o and λ_s constant then in T component hours we would expect $\lambda_o T$ open-circuit failures, $\lambda_s T$ short-circuit failures, and a total of $\lambda_o T + \lambda_s T = (\lambda_o + \lambda_s)T = \lambda T$ failures.

It should be noted that the component cannot be treated as two separate series elements (see Chapter 6) having failure rates of λ_o and λ_s, since failure into open-circuit (o/c) and short-circuit (s/c) are mutually exclusive events and thus cannot be independent (see Appendix 3).

Using probability theory it can be shown that:

$$\text{Probability of s/c failure} = \begin{array}{l}\text{Prob. of s/c failure}\\ \text{given that failure}\\ \text{has occurred}\end{array} \times \begin{array}{l}\text{Prob. of failure}\\ \text{of either s/c}\\ \text{or o/c}\end{array}$$

$$= [\lambda_s/\lambda]\,[1 - \exp(-\lambda t)]$$

*In some circumstances it may be necessary to use heating to prevent components becoming too cold, but it is more usual for high ambient temperatures to be a problem rather than low.

with a corresponding expression for the probability of o/c failure.

The total failure rate of the component can be broken down into percentage values of open-circuit $[100\lambda_o/(\lambda_o + \lambda_s)]$ and short-circuit $[100\lambda_s/(\lambda_o + \lambda_s)]$ failure rates. Some data on component failure modes is given in Smith (1985) and O'Connor (1985).

5.8.2 Competing Risks Model

It is useful to consider the competing risks model as providing the statistical data for reliability prediction. (A description of the model and further references are given in Mann *et al.* (1974)). Each failure mode is assumed to generate a separate randomly-derived time-to-failure; n failure modes would thus generate n times-to-failure, $t_1, t_2, \ldots, t_n$. The component then actually fails at the shortest of the n failure times.

Dr R. A. Evans (Evans (1980)) has pointed out that there is an apparent paradox in that failure into different modes apparently introduces events which are both mutually exclusive and independent (see Appendix 3). The resolution of the apparent paradox lies in understanding that the statistically independent competing processes carry on until the risk matures; the mutually exclusive states occur after the risk matures.

5.9 Design of Reliable Components

Since there are many different types of components, it is not possible to give very specific advice on how to design reliable components. However, the following are good general guidelines:

(a) use as large a safety factor (power, voltage or current) as is compatible with the proposed cost, size and weight of the component;
(b) base the design upon already well-established principles.

The second recommendation does not imply that no new devices should ever be developed! It indicates that high reliability is usually associated with mature processes (of design and manufacture). System designers who wish to use novel types of components must recognise that these may be initially very unreliable. It has been noticeable that undersea cable communication systems (which have a good reliability record) have almost always been designed using components manufactured by out-of-date technologies.

Integrated circuits are now so complex that it is very difficult to prove the correctness of their designs. The Royal Signals & Radar Establishment at Malvern has designed a microprocessor called VIPER which is claimed to be the world's first commercially-available microprocessor for which formal mathematical methods have proved the equivalence between the chip design and the specification.

5.10 The Manufacture of Reliable Components

Broad recommendations are:

(a) Use the most mature manufacturing processes available (see section 5.9).
(b) Devote as much quality control to the raw materials and to the manufacturing processes as is compatible with the component cost.
(c) Avoid contamination of the raw material and of the manufacturing processes as far as is economically possible. ('Clean room' techniques are used in the manufacture of the most reliable components.)

5.11 Component Testing

5.11.1 Destructive Testing

Data from the life testing of components can be analysed using the statistical methods described in Chapters 3 and 4. In order to save time, elevated stresses are often used and the results are extrapolated (using Arrhenius's equation) to normal stress and ambient conditions.

5.11.2 Non-destructive Testing

Non-destructive testing can be used by the component manufacturer to reject potentially unreliable components and thus improve the reliability of the components which are released to the customer. To be economically viable, the test must be cheap to run and easy to apply. A wide range of techniques can be used including X-rays and ultrasonics for mechanical tests. Some references are:

(a) Martensson, E. (1967). 'Non-linearity exposes bad electronic component parts', *Electronic Components*, April, pp. 371-373.
(b) Stanley, K. W. (1971). 'Improvements in reliability of metal film resistors', *Mullard Technical Communications*, No. 109, January, pp. 207-208.
(c) Kivenson, G. (1972). *Durability and Reliability in Engineering Design*', Pitman 1972. (The measurement of mechanical impedance is mentioned as a method of exposing potential unreliability.)
(d) *New Scientist* (1985). A series of articles on Non-destructive Testing in the magazine *New Scientist* beginning with the issue of 12 September.
(e) Chittick, R. C. and Gray, E. (1984). 'Improved Moisture Resistance of Multilayer Ceramic Capacitor Encapsulation by On-line Screening', *Proc 4th Capacitor and Resistor Technology Symposium*.

5.11.3 Self-testing

Complex VLSI chips can now have self-testing circuitry built-in to them. See Maunder, C. (1985). 'Built-in test: a review', *Electronics and Power (IEE)*, March, pp. 204–208.

5.12 Failure Rate Data

5.12.1 Sources

Some sources of constant failure rates are:

(1) US Dept of Defense: *MIL–HDBK–217*. (The number is followed by a letter to indicate the edition; the latest edition is 'F'.)
This is a very large (approximately 500 page) document which is not easy to use. However it is very comprehensive and allows for a selection from many different environments. Between publication of successive editions, updating notices are sometimes issued.
(2) *BT Handbook of Reliability Data* (HRD). (The latest edition is HRD5.)
(3) Bell System Information Publication: *Reliability Prediction Procedure for Electronic Equipment TR-TSY-000332*, Issue 1, September 1985.
Although this document is fairly long (68 pages) the table of failure rates is confined to 9 pages.
(4) Smith, D. J. (1985). *Reliability and Maintainability in Perspective*, Macmillan, 2nd edition.
This book not only contains a list of failure rates but also gives failure modes and the percentages of the total failure rate attributable to each mode.
(5) Centre National d'Etudes des Télécommunication (CNET) of France: *Recueil de Données de Fiabilité du CNET.* CNET failure rates are published in English as *Handbook of Reliability Data for Electronic Components RDF 93, English Issue 1993.*
(6) Nippon Telegraph & Telephone Public Corporation (NTT) of Japan: *Standard Reliability Table for Semiconductor Devices*.
(7) UK MoD: *DEF-STAN 00-41* (Part 3). This document gives failure rates for some mechanical and electrochemical components; the reader is referred to MIL-HDBK-217 for failure rates of electronic and electrical components.
(8) Rome Air Development Centre (RADC) of the USA: *Nonelectronic Parts Reliability Notebook* (NPRD). The latest edition is NPRD-3.

(9) Systems Reliability Service of the UKAEA: a databank is available for subscribers only.

(10) Det Norske Veritas (DNV): the OREDA databank is available to subscribers only.

There is very little published data on components used in very high reliability systems (such as undersea telephone cables). When required, the information must be obtained from the component manufacturers. It should be noted that the above published sources can give widely varying failure rates for the same component operating under supposedly identical conditions. The variation between the sources is disturbing (see Spencer (1986) and Smith (1986)) since factors of 10 or more can be involved (and it is claimed in each source that the data are based at least partly upon field failure data). Because of the differences between the various sources, it is vital that when any reliability prediction is made, the source of the failure rates being used should be clearly identified.

Quoted failure rates are usually based upon the 60 per cent upper confidence limit derived from the available data. This has been deemed to be the optimum method for providing data for reliability prediction. However, the interpretation of failure data (particularly field data) is not always easy and this may go some way in explaining the large discrepancies between failure data which has been noted above. Some problems in dealing with failure data are:

(a) If early failures are to be excluded, there may be uncertainty about whether a particular failure should be counted or not.

(b) Since equipment is usually introduced into the field over a period of time, there may be uncertainty about how many component-hours have been generated.

(c) If a repairman incorrectly removes a 'good' component before identifying correctly the failed component, the 'good' component may have to be replaced anyway (since it may have been damaged by extraction from the printed circuit board). A subsequent count of replaced components will then include the 'good' component even though it did not fail.

(d) When printed circuit boards are returned for repair, a fault sometimes cannot be identified ('fault not found'). There is then some uncertainty whether there is an accountable failure (caused by an intermittent fault or excessive parameter drift) or not.

In addition to the above quoted sources of failure rate data, most large electrical and electronic engineering companies keep their own failure databanks; this information is usually regarded as 'company confidential' and is not available for general use. There is an ongoing co-operative project between Loughborough University, the Ministry of Defence (UK) and some large companies on the collection and pooling of failure data.

5.12.2 The Use of Component Failure Rate Sources

Most sources of component failure rates use formulae for the calculation of a required rate. For uncomplicated components like resistors or capacitors, the formula is of the form:

$$\lambda = \lambda_b \pi_Q \pi_E$$

where λ_b is the base failure rate for the component type while π_Q and π_E are parameters whose values depend upon the component quality level and the system environment respectively. It is assumed that the component is stressed well within its capability, so that the failure rate is not strongly dependent upon the stress level (power, voltage or current) that is applied.

Complex components have formulae which involve more factors. As an example, the formula for Microelectronic Devices in MIL-HDBK-217E is

$$\lambda_p = \pi_Q \pi_L [C_1 \pi_T \pi_V + (C_2 + C_3) \pi_E]$$

where C_1, C_2 and C_3 are various complexity factors: π_V is a voltage derating stress factor; π_L is a learning factor depending upon the maturity of the manufacturing process; π_T is a temperature factor based upon the maximum junction temperature within the device; π_E is the environmental factor.

Three environments are defined in HRD4: Ground Benign (GB), Ground Fixed (GF) and Ground Mobile (GM). No less than 20 different environments are defined in MIL-HDBK-217E: Ground, Benign; Ground, Fixed; Ground, Mobile; Space, Flight; Manpack; Naval, Sheltered; Naval, Unsheltered; Naval, Undersea, Unsheltered; Naval, Submarine; Naval, Hydrofoil; Airborne, Inhabited, Transport; Airborne, Inhabited, Fighter; Airborne, Uninhabited, Transport; Airborne, Uninhabited, Fighter; Airborne, Rotary, Winged; Missile, Launch; Cannon, Launch; Missile, Free Flight; Airbreathing Missile (Flight). It is not clear to the author how the US authorities have enough failure data adequately to differentiate between failure rates for all these different environments. In HRD4, π_E varies between 1.0 (Ground, Benign) and 8.0 (Ground, Mobile).

Some devices are required to operate once only (e.g. fuses), or are unlikely to operate frequently (e.g. changeover switches); these are called one-shot devices and are sometimes given a time-dependent probability that they will operate successfully when required to do so.

5.12.3 Parts Count and Parts Stress Analyses

See section 6.10.

5.13 Databases on Computers

Databases holding failure rate data are available for personal computers, minicomputers and mainframes. The most widely used are those for personal or minicomputers; examples of programs holding such data are:

HardStress (HRD data) from Item Software (Fareham, Hants., UK)
FRATE (HRD data) from BT
MilStress (MIL-HDBK-217 data) from Item Software; this program can also extract failure rates from a library of NPRD data

These and similar programs also have the facility for making system failure rate calculations (see Chapter 19).

An example of a database on a mainframe computer is the CODUS databank which carries a wide variety of component information including HRD and MIL-HDBK-217 failure rate data (Codus Ltd, Sheffield, UK).

5.14 Effect of Change of Environment or Usage

See section 6.11.

5.15 Standards

Standards which contain reliability-related material are published by:

British Standards Institution	- BS series
US Department of Defense	- MIL-STD series
International Electrochemical Committee	- IEC publications
UK Ministry of Defence	- DEF-STD series
Cenelec Electronic Components Committee	- CECC series
German Standards Institution	- DIN series
British Telecom	

Cenelec is an abbreviation for Comité Européen de Normalisation Electrotechnique.

6 System Reliability

6.1 Introduction

During system development it is essential that reliability predictions should be made at all stages of the development in order to ensure that the finished product meets its specification (product objectives). A first step in the reliability prediction of any system is the identification of failure modes; a reliability analysis must then be carried out on each one separately. In this chapter a discussion of failure modes is followed by methods of predicting the reliability of the simplest type of system, the series reliability system, using component failure data. Chapter 7 briefly describes methods of increasing system reliability and then Chapters 8 and 9 give reliability prediction methods for the more complex fault-tolerant systems.

In order to simplify the concepts involved, the repair of the systems under consideration is left to be considered at a later stage, see Chapters 10, 11 and 14.

It is shown that if a system can be considered to be in 'series' reliability and is constructed of constant failure rate components, then the system itself has a constant failure rate.

Sections 6.3 to 6.7 are concerned with series systems; sections 6.8 onwards deal with more general topics in system reliability.

6.2 System Failure Modes

It is not always easy to decide on failure criteria. As an example, consider the failure criteria of a car. If it will not move under its own power then it has clearly failed. But should the failure of the windscreen wipers also be accounted a total failure? Although the car may still be usable in fine weather, it may be totally unusable in heavy rain.

There are thus categories of failure, the most serious being total failure. Other, less serious, modes of failure are those in which a degraded service is provided.

As a further example consider a stereo radio cassette recorder. This will have at least the following modes:

Failure of the radio only
Failure of cassette player or recorder only
Complete failure

A reliability analysis of each mode of failure is possible. In this and succeeding chapters, methods of reliability analysis are presented and reference to a 'system' should be understood to refer not only to a simple system with only one failure mode but also possibly to a more complex system for which the analysis is of a particular failure mode.

Systems which contain software are liable to have failures caused by the appearance of software 'bugs' as well as those caused by component failures. Since these failures are of two distinctly different types, they will be treated separately. In Chapters 6 to 12, only hardware failures are considered: software reliability is discussed in Chapter 13.

6.3 Series Reliability: Reliability Block Diagrams

A number of components, connected together so as to form a system, are said to be in series reliability if the failure of any one component causes the failure of the total system. (The system can be regarded as being 'fault-intolerant'.) Series reliability is shown diagrammatically in Figure 6.1 which is called a Reliability Block Diagram (RBD). It must be emphasised that such diagrams do not represent the way that the components are interconnected physically, but show how they must be treated from a reliability viewpoint. It is quite possible for two components to be connected physically in parallel (a resistor and a capacitor, for instance) but to be in series reliability (so long as failure of either causes system failure). Equally, two components may be physically in series but in parallel from a reliability viewpoint. (Parallel reliability is introduced below.)

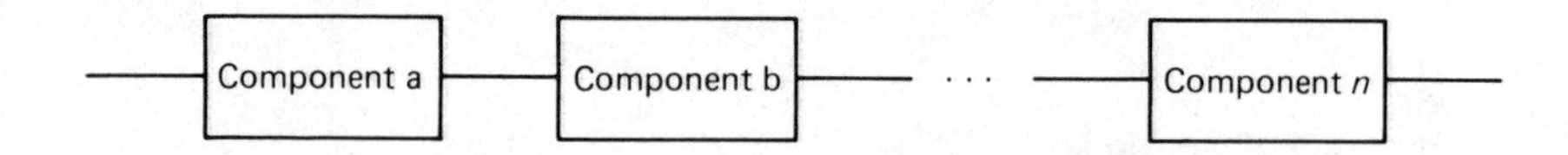

Figure 6.1 Reliability block diagram for a series system

In simple RBDs like Figure 6.1, arrows are usually omitted since the meaning of the diagram is quite clear. In more complex RBDs (see, for example, Figure 9.1) it is necessary to add arrows to the diagram in order to make the meaning clear.

If the probabilities of failure of the n components in Figure 6.1 are independent then we can consider the probability of a component's failing inside the system to be the same as the probability of its failing outside the system, so long as it is tested at ambient conditions and stress levels outside the system which are identical with those which it experiences inside the system.*

Thus we can predict the reliability of the system if we have suitable reliability information on all of its components. If the probabilities of failure of the components can be considered to be independent, then the system reliability $R_s(t)$ is the product of the n component reliabilities:

$$R_s(t) = R_a(t)\, R_b(t) \ldots R_n(t)$$

In probabalistic terms this is equivalent to:

Probability of system survival	=	Probability of survival of component a	*and*	Probability of survival of component b	*and . . . and*	Probability of survival of component n

It should be noted especially that the probability of system failure is *not*:

$$F_a + F_b + \ldots + F_n$$

The fallacy is most easily shown by considering the Venn diagram for two components a and b only; this is shown in Figure 6.2.

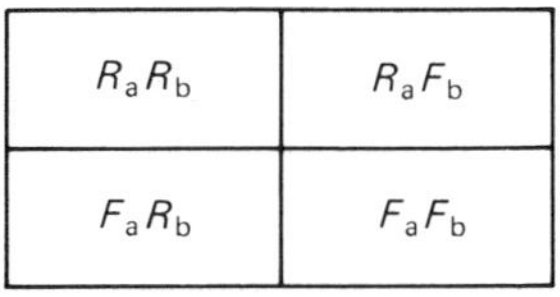

R_aR_b	R_aF_b
F_aR_b	F_aF_b

Figure 6.2 Venn diagram for probabilities of survival and failure of two components connected in series

It is readily seen from the figure that

Probability of system failure $= R_aF_b + F_aR_b + F_aF_b = F_a + F_b - F_aF_b$

This is the correct expression for

Probability of system failure	=	Probability of failure of component a	*or*	Probability of failure of component b

*In principle, this is what is required. In practice, components are tested under particular stresses and environmental conditions. Failure rates for other stresses and environmental conditions are found by extrapolation.

since the probabilities of failure of a and b are assumed to be independent. For n components connected in series, it can be shown by expanding

$$F_s = 1 - R_a R_b \ldots R_n = 1 - (1 - F_a)(1 - F_b) \ldots (1 - F_n)$$

that

$$F_s = \begin{array}{c}\text{Sum of } F \text{ terms} \\ \text{taken one at a} \\ \text{time}\end{array} - \begin{array}{c}\text{Sum of } F \text{ terms} \\ \text{taken two at a} \\ \text{time}\end{array} + \begin{array}{c}\text{Sum of } F \text{ terms} \\ \text{taken three at a} \\ \text{time}\end{array} - (\text{etc.})$$

Then for three components connected in series:

$$F_s = (F_a + F_b + F_c) - (F_a F_b + F_b F_c + F_c F_a) + F_a F_b F_c$$

As an example of this form of analysis, consider a system which consists of the circuit shown in Figure 6.3(a). So long as

(a) failure of any component causes complete circuit failure
(b) the probabilities of failure of the components are independent

then the reliability block diagram of this system can be represented by Figure 6.3(b). (Failure of the DC supply is neglected.) The system reliability is

$$R_s(t) = R_a(t)\, R_b(t)\, R_c(t)\, R_d(t)\, R_e(t)\, R_f(t)$$

That this is a reasonable procedure can be seen by examination of Figure 6.4, which represents a system having three components a, b and c. It is assumed that the component reliabilities are known for a mission time t_M and that

$$R_a(t_M) = 0.8 \quad R_b(t_M) = 0.6 \quad R_c(t_M) = 0.75$$

Figure 6.3 then represents how, on average, 100 systems would behave. The columns represent the reliabilities of the three components and each should be regarded as being divided into 100 rows, each row containing one each of a, b and c, and thus representing a system. The average numbers of failed components are shown by the shaded areas.

It can be seen that out of 100 systems, only 36 have components a, b and c working at t_M and 36/100 can be equated with

$$R_a(t_M)\, R_b(t_M)\, R_c(t_M) = 0.8 \times 0.6 \times 0.75 = 0.36$$

If for ease of nomenclature we put

$$R_a(t_M) = a$$

and

$$F_a(t_M) = 1 - R_a(t_M) = \overline{a}$$

then the possible system states may be identified with the expansion of $(a + \overline{a})$ $(b + \overline{b})\,(c + \overline{c})$ and we see that the eight possible system states are:

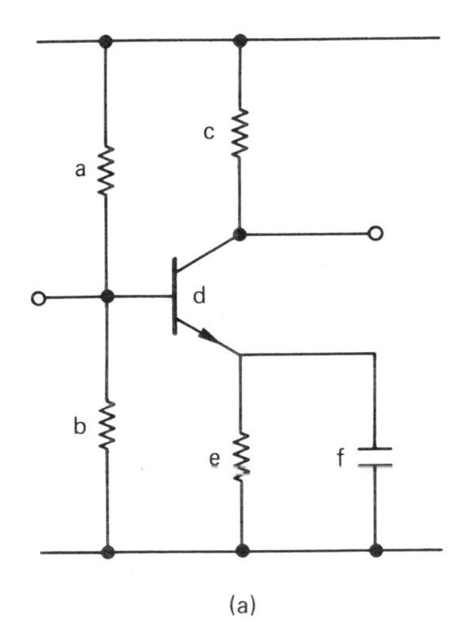

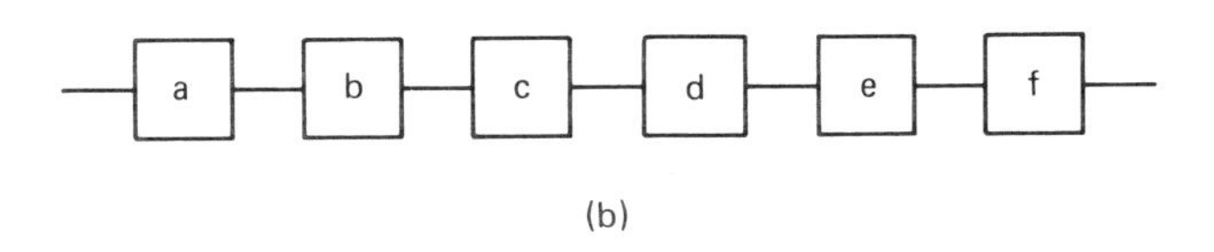

Figure 6.3 A simple circuit and its series reliability block diagram: (a) circuit; (b) reliability block diagram (RBD)

Up state $\quad a\,b\,c$

Down states

$$\left\{\begin{array}{ll} \bar{a}\,b\,c & \bar{a}\,b\,\bar{c} \\ a\,\bar{b}\,c & a\,\bar{b}\,\bar{c} \\ a\,b\,\bar{c} & \bar{a}\,\bar{b}\,\bar{c} \\ \bar{a}\,\bar{b}\,c & \end{array}\right.$$

These are the 2^3 terms obtained from the expansion of $(a + \bar{a})\,(b + \bar{b})\,(c + \bar{c})$. If there are n components, the possible number of states is 2^n.

The probability of the system being in any one of the above states can be equated with the corresponding proportion of systems from Figure 6.3; for example

$$a\,b\,\bar{c} = 0.8 \times 0.6 \times 0.25 = 0.12 = 12/100$$

This example is somewhat artificial in that once the first component of a series system has failed, the analysis is not usually continued any further because the system, as a whole, is no longer in a working state. (The case of redundancy, in which failure of a particular component or unit may not cause total system failure, is considered below.)

It should be noted how quickly the reliability of a system declines with the number of components used, as for example:

$$(0.9)^3 = 0.729$$

$$(0.9)^4 = 0.656$$

$$(0.9)^5 = 0.590$$

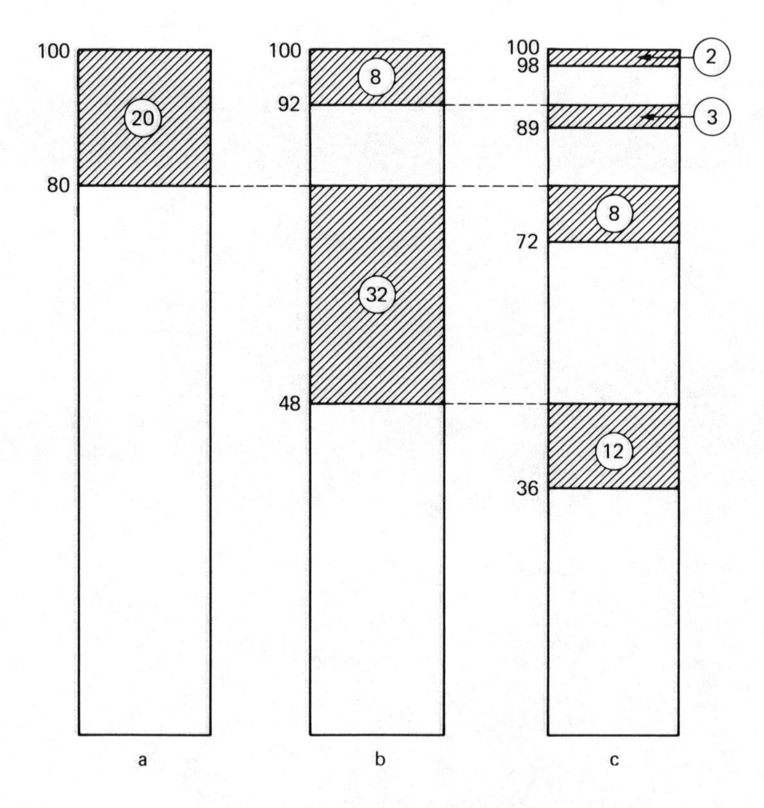

Figure 6.4 Illustration of system states

6.4 Series Reliability of Subsystems, Units and Modules

The 'system' of Figure 6.3(a) is a very simple system indeed. Most systems can be considered to be a collection of subsystems which, depending upon their complexities, can be further broken down into units, modules (usually printed circuit boards) and eventually individual components. The RBD of subsystems for a particular failure mode can usually be derived from the functional block diagram of the system.

As an example, Figure 6.5 shows the functional block diagram of the stereo cassette radio player whose failure modes were introduced in section 6.2.

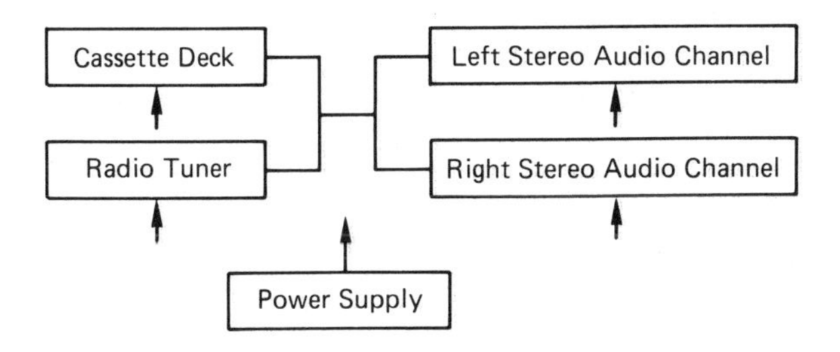

Figure 6.5 Functional block diagram of a stereo cassette radio player

The four failure modes will have the RBDs shown in Figure 6.6. It should be noted that for the first two failure modes, the two audio stereo channels have been placed in series; this corresponds to the failure definitions of section 6.2 since failure of either stereo channel is counted as contributing to that particular failure mode.

Subsystems are shown in parallel for the last failure mode ('complete failure') because it needs both audio channels or both the radio tuner and the cassette deck to fail in order to cause total system failure. Subsystems which are in parallel from a reliability viewpoint are considered in Chapters 8 and 9.

In probabalistic terms, the 'complete failure' block diagram of Figure 6.6 can be interpreted as

Probability of total failure	=	Probability of failure of tuner *and* deck	*or*	Probability of failure of left *and* right channels	*or*	Probability of failure of power supply

6.5 General Form of the Reliability Function for Series Reliability

It has been considered up to now that the reliability of a component is known at some mission time t_M. This value t_M may obviously be regarded as a dummy variable and replaced by an unspecified time so that the general formula for the reliability of a system consisting of n components in series is

$$R_s(t) = R_a(t)\, R_b(t) \ldots R_n(t)$$

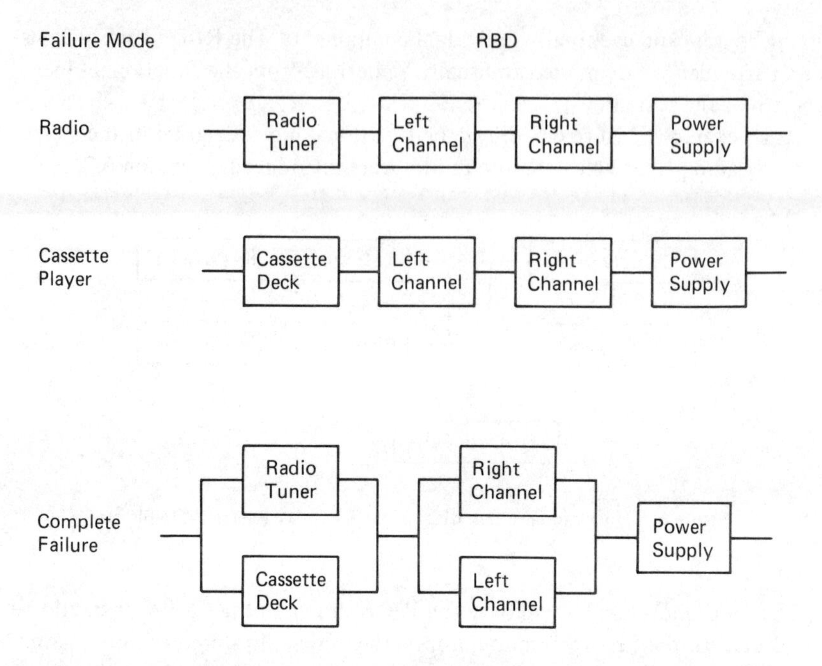

Figure 6.6 RBDs for the failure modes of a radio cassette stereo player

So long as the probabilities of failure are independent (an assumption which is usually valid), this expression can be used whether all components are CFR or not. It is shown in the next section that if all the components have CFRs then the whole system has a CFR. If this is not so, then analytical or numerical methods can be used to calculate $R_s(t)$. If necessary, the system MTTF can be found by numerical computation of the area under the $R_s(t)$ curve.

6.6 Importance of Constant Failure Rate in Series Reliability

Consider a series system of n components, all of which have constant failure rates. Then

$$R_a(t) = \exp(-\lambda_a t),\quad R_b(t) = \exp(-\lambda_b t), \ldots,\quad R_n(t) = \exp(-\lambda_n t)$$

so that the system reliability is

$$\begin{aligned} R_s(t) = R_a R_b \ldots R_n &= \exp[-(\lambda_a + \lambda_b + \ldots + \lambda_n)t] \\ &= \exp[-\lambda_s t] \end{aligned}$$

where $\lambda_s = \lambda_a + \lambda_b + \ldots + \lambda_n$.

It follows that the system MTTF, denoted by θ, is

$$\theta_s = \frac{1}{\lambda_s} = \frac{1}{\lambda_a + \lambda_b + \ldots + \lambda_n}$$

Owing to the fact that the exponential distribution is characterised by the single parameter λ, it follows that any desired reliability parameter (MTTF, Probability of survival for time T etc.) can be calculated for the system from the knowledge of λ_s.

6.7 Assumptions Made in Making a Series Reliability Prediction by Adding Failure Rates

(a) The components are constant failure rate devices (or may be treated as such).
(b) The components are inside their useful life.
(c) The components are substantially homogeneous with those whose failure rates have been measured.
(d) The system is manufactured according to accepted practices.
(e) When placed in the system, the components have the same ambient conditions and stress levels as those under which the failure rates were measured (or calculated by extrapolation from measured data).
(f) The component failure probabilities are independent.
(g) The design of the system is satisfactory.
(h) The system is not misused.

6.8 Series Reliability Systems - General Formula

It has been shown that for a particular component

$$R = \exp\left[-\int_0^t \lambda(u)\,du\right]$$

Thus in a series reliability system containing n components, the system reliability $R_s(t)$ is given by

$$R_s(t) = R_1(t)\,R_2(t)\ldots R_n(t)$$

($R_i(t)$ is the reliability of the ith component); it follows that

$$R_s(t) = \exp\left[-\int_0^t [\lambda_1(u) + \lambda_2(u) + \ldots + \lambda_n(u)]\,du\right]$$

6.9 System Failure Modes from Component Failure Modes

Knowledge of component failure modes (and their relative frequencies) can be used to predict system failure modes (see section 19.3 on Failure Mode, Effect and Criticality Analysis) and system unreliability. The method is best illustrated by a simple example.

Consider two switches a and b connected in series as shown in Figure 6.7. Assume that the switches can fail only into open-circuit and short-circuit. Let the reliability of switch i be R_i; the probability of failure into open-circuit be F_{io}; the probability of failure into short-circuit be F_{is}.

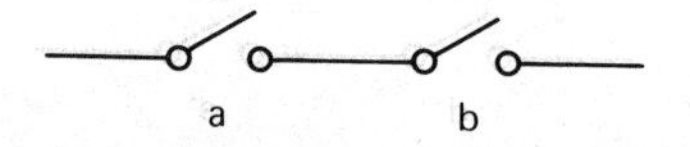

Figure 6.7 Two switches physically connected in series

Let us consider that the system is operational if one or both switches can be used to open or close the connection. There are then two failure modes: (a) the connection is permanently short-circuit; (b) the connection is permanently open-circuit.

We can put

$$(R_a + F_{ao} + F_{as})(R_b + F_{bo} + F_{bs}) = 1$$

since the terms inside each bracket sum to unity. Expansion of the left-hand side of the equation gives all possible combinations of states of the switches. The nine terms can be segregated into three groups which correspond to the operational mode and two failure modes. The first group which sums to R_s (the system reliability) represents the probability that the system is operational:

$$R_s = R_a R_b + R_a F_{bs} + F_{as} R_b$$

The second group represents the probability that the system has failed into short circuit (F_{ss}):

$$F_{ss} = F_{as} F_{bs}$$

The third group represents the probability that the system has failed into open circuit (F_{so}):

$$F_{so} = R_a F_{bo} + F_{ao} R_b + F_{ao} F_{bo} + F_{ao} F_{bs} + F_{as} F_{bo}$$

If the two switches have constant failure rates and the relative percentages of failure into open-circuit and short-circuit are known, then using the nomenclature of section 5.8 it is possible to put

$$R_a = \exp(-\lambda_a t)$$

$$F_{ao} = [\lambda_{ao}/\lambda_a]\,[1 - \exp(-\lambda_a t)]$$

$$F_{as} = [\lambda_{as}/\lambda_a]\,[1 - \exp(-\lambda_a t)]$$

with corresponding expressions for R_b, F_{bo} and F_{bs}.

From these expressions, it is possible to calculate the desired reliability parameters.

6.10 Parts Count and Parts Stress Analyses

It is usual to distinguish between Parts Count and Parts Stress Analyses although in practice the distinction between them may not be sharply defined.

In Parts Count Analysis, sums are made of how many components of each type contribute to a particular failure mode and a failure rate is allocated to each component type without going into details of the stress levels of each individual component. The total failure rate is then obtained by summing all the products of the form: (number of resistors) × (resistor failure rate).

Parts Stress Analysis is similar to the above except that the stresses on each individual component are assessed and a corresponding failure rate assigned on a component-by-component basis.

An example of a parts count analysis is shown in Figure 6.8 for the circuit depicted in Figure 6.3(a). The analysis sheet includes a column for the percentage contribution of each type of component to the total failure rate. This information can be very useful for determining the sensitivity of a reliability prediction to assumptions made regarding component failure rates; it is also obviously useful in showing which are the most important components to change if the failure rate of the module should be reduced.

The tabulation of the results in Figure 6.8 is suitable for HRD and Bell failure rates sources in which the environment factor π_E modifies the summed failure rate. It would be necessary to add an extra column for π_E if MIL-HDBK-217 failure rates were used because then the formula using π_E varies from component to component.

In the simple example shown in Figure 6.8 it has not been necessary to segregate the components into categories, such as 'traffic-affecting' and 'non-traffic-affecting'. In more complex analyses it may be useful to add an extra column to the figure so that each component may have its category listed. A more realistic example of the addition of failure rates is shown in Figure 6.9 which represents the calculation of the failure rate for a printed circuit board used in a telecommunications application.

RELIABILITY PREDICTION OF:

SYSTEM: MB
SUBSYSTEM: A
UNIT: A10
MODULE: Amplifier

DATE: 6 Oct. 1988

ENVIRONMENT: Ground, Benign

QUALITY LEVEL: HRD4 Level 2

Component reference	Component type	No. used (a)	Component failure rate (fits) (b)	Quality factor π_Q (c)	Comp. total failure rate (a x b x c)	Percent contribution button to total	Remarks
a, b, c, e	Resistor, Carbon composition	4	0.8	1.0	3.2	2.6	
d	Transistor Silicon Bipolar Power	1	100	1.0	100	81.2	
f	Capacitor, Al Electrolytic Foil	1	20	1.0	20	16.2	
				TOTAL (λ_T)	123.2		

VALUE OF ENVIRONMENTAL FACTOR (π_E) = 1.0

MODULE FAILURE RATE $= \pi_E \times \lambda_T$
$= 1.0 \times 123.2$
$= 123.2$ fits

Figure 6.8 Example of parts count analysis for the circuit of Figure 6.3(a)

RELIABILITY PREDICTION OF:

SYSTEM: PMDC
SUBSYSTEM: B
UNIT: B1
MODULE: SIGNAL PROCESSOR
SOURCE OF FAILURE RATE DATA: HRD3

DATE: August 1986
ENVIRONMENT: GB
QUALITY LEVEL: Various

Component type	No. used (a)	Component failure rate (fits) (b)	Quality factor π_Q (c)	Component total failure rate (a × b × c)	Percent contribution to total	Remarks
Capacitor	35	0.2	1.0	7	0.1	
Capacitor (polarised)	1	1	1.0	1	0.0	
74 Series Logic IC	26	50	1.0	1300	27.0	Critical
SMC ULA	2	600	1.0	1200	24.9	Bus I/F
PROM	1	50	1.0	50	1.0	Bipolar
EPROM	3	30	1.0	90	1.9	MOS 32K × 8
SRAM	5	15	1.0	75	1.6	MOS 256 × 4
Programmed PROM	7	70	1.0	490	10.2	
SRAM	1	70	1.0	70	1.5	2K × 8
Resistor Network	2	15	1.0	30	0.6	
Resistor	22	0.2	1.0	4.4	0.1	
Toggle Switch (DIL)	1	30	1.0	30	0.6	
Xtal Controlled OSC	1	90	1.0	90	1.9	
Connectors	178	0.7	1.0	125	2.6	
PCB	1	30	1.0	30	0.6	
74 Series Logic IC	24	50	1.0	1200	24.9	Critical
74 Series Logic	3	10	1.0	30	0.6	
			TOTAL (λ_T)	4822		

VALUE OF ENVIRONMENTAL FACTOR (π_E) = 1.0

MODULE FAILURE RATE = $\pi_E \times \lambda_T$
= 1.0 × 4822
= 4822 fits

Figure 6.9 Parts count analysis for a printed circuit board

6.11 Effect of Changing Environment or Use

During the design life of a system, the environment in which it operates or the use to which it is put may change. These changes are revealed in the system's mission profile as illustrated in Figure 2.3. The changes may be cyclic (say a variation of ambient temperature between day and night), or non-cyclic (say a period of storage followed by permanent use).

In the case of non-cyclic changes, consider that there is only one change from an initial condition in which the failure rate is $\lambda_1(t)$ to a final condition in which it is $\lambda_2(t)$. (The treatment is easily extended to deal with more than one change.)

A general expression is

$$R(t) = \exp\left[-\int_0^{t_1} \lambda_1(t)\mathrm{d}t + \int_{t_1}^{t} \lambda_2(t)\,\mathrm{d}t\right] \quad \text{for } t > t_1$$

When the failure rates (λ_i) during the two stages ($i = 1, i = 2$) are constant, then it is easily shown that

$$\theta = 1/\lambda_1 - \exp(-\lambda_1 t_1)/\lambda_1 + \exp(-\lambda_2 t_1)/\lambda_2$$

When the failure rates are non-constant, then $R(t)$ must be evaluated using the appropriate failure rate functions.

In the case of cyclic changes, we shall for simplicity consider only two conditions which last alternately for times t_1 and t_2.

In the constant failure rate case, it is readily seen that after a large number of cycles the system reliability function $R_s(t)$ can be approximated* by:

$$R_s(t) = \exp(-\lambda_m t)$$

where λ_m is the mean failure rate:

$$(\lambda_1 t_1 + \lambda_2 t_2)/(t_1 + t_2)$$

Hence the MTTF is approximately $1/\lambda_m$.

In the non-constant failure rate case, valid approximations depend upon the forms of the failure rates during the appropriate periods.

6.12 Expected Number of Failures in a Given Time

The treatment of this problem for systems is the same as that for components which has been given in section 3.16.

*Since $R(t) = \exp[-\int_0^t \lambda(u)\mathrm{d}u]$.

6.13 Statistical Treatment of System Failure Data

Raw failure data may be plotted on Weibull Probability Paper as shown in sections 4.5 and 4.6 in order to establish whether or not the failure rate is constant. If this is so, then point estimates and confidence intervals may be calculated using the methods of Chapter 4. If the failure rate is not constant, then an attempt should be made to see whether the Weibull plot can be shown to consist of separate portions (for example, a decreasing failure rate followed by a constant failure rate).

6.14 Composite Failure Pattern

See section 3.17.

Exercises

6.1. Consider a series reliability system consisting of 100 components each having a constant failure rate of λ.

(a) If $\lambda = 600 \times 10^{-9}$/hour, what is the system reliability for a mission time of 1000 hours?

(b) If the required reliability of the system over a mission time of 1000 hours is 0.98, what value of λ would be required to meet this target?

(c) If $\lambda = 600 \times 10^{-9}$/hour, what mission time is allowable for a target reliability of 0.98?

6.2. A flip-flop circuit is shown in Figure 6.10. Constant failure rates (in units of 10^{-9}/hour (fits)) for the components and solder joints at 25 °C and 85 °C are as follows:

Component	*Temperature* 25 °C	85 °C
Resistor	0.2	2
Capacitor	2	16
Transistor	20	44
Solder joint	0.07	0.07

What is the MTTF of this circuit (neglecting the power supply) at ambient temperatures of 25 °C and 85 °C?

What is the probability of survival at each temperature for a mission time of 8000 hours assuming that failure of any component causes system failure?

(Assume that resistors and capacitors each require two solder joints and that transistors each require three solder joints.)

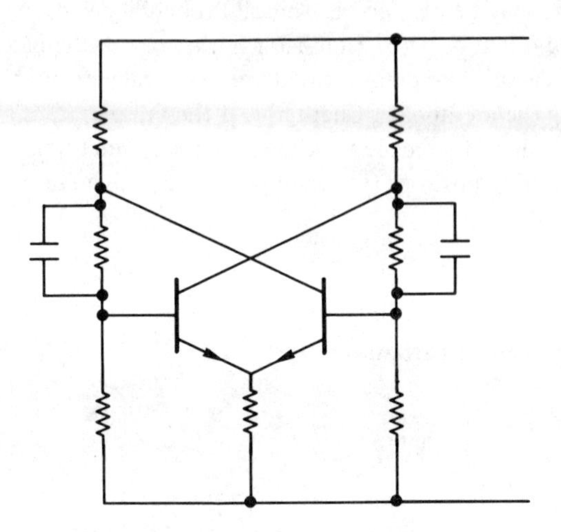

Figure 6.10 A flip-flop circuit

6.3. A certain unit is constructed from five component types. The component types, numbers used and failure rates (per component) are as follows:

Component Type	*Number Used*	*Component Constant Failure Rate (fits)*
a	3	10
b	4	10000
c	10	1000
d	7	850
e	2	2000

Calculate the unit failure rate and its MTTF (neglecting the contributions of the printed circuit board, solder joints and connectors).
Calculate the percentage contribution to the total of each component type.
If the failure rate of component b could be reduced to 5000 fits, what would the new MTTF of the unit become? What would the percentage contribution of component type b to the total failure rate become?

7 Methods of Increasing System Reliability

7.1 Introduction

The RBD for the 'total failure' mode in a newly-designed system is likely to be a series connection of subsystems each of which consists of a series of units and modules. The overall failure rate must be divided between the subsystems, and preliminary predictions made to assess whether or not the system is likely to meet its reliability target; reliability allocation methods are described in Chapter 16. RBDs for other failure modes (if they are highly critical) may also have to be constructed. In cases where the preliminary designs seem unlikely to meet the specification, then methods of increasing system reliability must be sought.

7.2 Methods

Methods of improving system reliability are:

(a) *During the Design Phase*

(i) Use fewer components; for example, by (1) simplifying the system;
or (2) using more complex (possibly custom-designed) integrated circuits.

(ii) Use better components, that is, (a) better quality;
and/or (b) more highly derated.

(iii) Improve the environment, for example, use cooling fans, reduce vibration, etc. (see section 5.6.3).

(iv) Use fault-tolerance in the system (see Chapters 8 and 9 for Hardware; Chapter 13 for Software).

(v) Use (or improve) preventive and/or corrective maintenance.

(b) *During the Development, Pre-Production and Production Phases*

In addition to the above, the following methods may be used:

(vi) Improve the design (for example, to allow increased drift in component values).

(vii) Burn-in the components and/or system.

(viii) Use non-destructive testing of components (see section 5.11.2).

(ix) Improve QC during manufacture by:
(1) improving assembly methods;
(2) improving inspection criteria.
(x) Improve maintenance techniques (for example, by (1) using better test equipment and/or (2) early replacement of limited-life components).
(xi) Improve software reliability (see Chapter 13).

The effects of methods (i) to (iii) have already been described in Chapter 5 while the advantages (and disadvantages!) of burn-in and non-destructive testing are treated in sections 5.5.1 and 5.11.1. The next two chapters describe methods of analysing fault-tolerant systems (method (iv) above). Methods (v) and (x) depend upon maintenance (Chapters 10 and 14) while software reliability is discussed briefly in Chapter 13. The remaining methods for increasing system reliability are treated in Chapter 16.

7.3 Comments

7.3.1 System Simplification

It may not always be possible to simplify a system so as to increase its reliability. Nevertheless it is usually worthwhile to pose the question 'can the system be simplified?', because it is quite common for over-elaborate system requirements to be specified at the initial design stage. The question 'do you really need an all-singing and all-dancing system?' can encourage a design team to re-consider whether an ultra-sophisticated system is really needed. Designers of military systems may well have learnt to shudder on hearing the words 'wouldn't it be nice if . . .'.

7.3.2 Decrease in Component Count

An integrated circuit usually has a lower failure rate than the group of discrete components which it replaces, and so system reliability can often be improved by replacing a circuit using many discrete components by a single integrated circuit. Custom-built integrated circuits are usually expensive and so their use will usually involve a cost penalty.

7.3.3 Use of Fault Tolerance

A series system as described in sections 6.3 to 6.5 can be described as 'fault intolerant' because failure of any one component will cause total system failure. A fault-tolerant system is one in which at least some parts of the system may fail without causing total system failure. An example of fault-tolerance is in the design of a two-engined aircraft which is capable of flying on one engine. However, there is only limited fault-tolerance in the aircraft; a major structural failure (for example, loss of the tailplane) will still cause the aircraft to crash.

7.3.4 Use of Preventive Maintenance

Preventive maintenance is fully defined in Chapter 10. Briefly, it is maintenance aimed at preventing failures and is exemplified by the regular maintenance actions which are taken with cars, like checking tyre pressures, checking oil and coolant levels etc. It may not be easy to decide during the Design Phase on exactly what impact preventive maintenance will have, although the early replacement of limited-life components brings an obvious improvement to reliability. During the Production Phase a careful study and analysis of the failures which occur in the field may indicate how, when and where preventive maintenance should be introduced or extended.

7.3.5 Use of Corrective Maintenance

Corrective maintenance (repair) is described in detail in Chapter 10; it consists of those actions which return a failed system to working order.

From a reliability viewpoint, corrective maintenance chiefly affects the system availability (see section 10.2), and so faster and more effective corrective maintenance will generally decrease the system down time and increase the system availability.

7.3.6 Design Improvements

It is sometimes suggested that potential sources of failure should be 'designed-out'. For example, stringent testing of a system might show that the drift with temperature in the current gain of a transistor is large enough for the system output to fail to meet its specified level. Re-design of the circuit may reduce the criticality of the current gain and so make the system more reliable. However, 'designing-out' of failure sources is generally possible to only a limited extent and this is why fault-tolerance is often used, particularly in systems where failure can cause a safety hazard (for example, in the control systems of nuclear reactors). Environmental testing may highlight the need for protection against corrosion or fungus attack.

7.3.7 *Quality Control*

The importance of Quality Control during system manufacture is stressed in Chapter 16. It is only necessary to remark here that any relaxation in Quality Control (for example, by allowing a solder bath to become contaminated) can have a catastrophic effect on an otherwise reliable system.

7.3.8 *Cost*

Improvement to system reliability will almost without exception cause increase to development costs, system cost (to the customer) and maintenance cost (to the manufacturer and/or customer). Although the reliability engineer can do little wholly to prevent these increases, what he can do is to minimise them as far as possible. For example, with constantly increasing cost of corrective maintenance (repair), it may be worthwhile to spend more money on fault-identification and automatic alarms rather than on increasing the number of maintenance staff.

8 Fault-tolerant Systems – I

8.1 Introduction: Fault Tolerance

The series reliability systems considered in Chapter 6 are 'fault-intolerant' because failure of any component causes system failure. In this and the succeeding chapter we consider systems in which some components may fail without causing system failure; they are thus 'fault-tolerant' and to achieve this some measure of hardware redundancy must be introduced. (Fault-tolerance in software is considered in Chapter 13.)

There are two main types of redundancy: active and passive (or standby) redundancy.

In active redundancy two or more units are placed in parallel reliability. Normally the load is shared but the units are such that if one unit (or possibly more) fails then the remainder can carry the full load. (An example of active redundancy is a two-engined aircraft which is capable of flying on one engine.) It must be emphasised that if two or more units are in parallel reliability, they are not necessarily connected together in parallel physically.

In passive redundancy, one unit normally carries the full load but if that unit should fail then another (not necessarily identical) unit is switched in to take its place; provision may be made for yet more standby units.

8.2 Active Redundancy – Parallel Reliability

Consider two units a and b connected in parallel reliability as shown in Figure 8.1.

The system fails only when both a and b have failed: if the probabilities of failure are independent, then the system failure function F_s is given by

$$\begin{aligned} F_s(t) &= F_a(t)\, F_b(t) \\ &= [1 - R_a(t)]\,[1 - R_b(t)] \\ &= 1 - R_a(t) - R_b(t) + R_a(t)\, R_b(t) \end{aligned}$$

where $R_a(F_a)$, $R_b(F_b)$ are the reliability (failure) functions of units a and b respectively. Hence

$$R_s(t) = 1 - F_s(t) = R_a(t) + R_b(t) - R_a(t)\,R_b(t)$$

The expression for $R_s(t)$ can also be found by expanding the expression $[R_a(t) + F_a(t)]\ [R_b(t) + F_b(t)]$ and then picking out those terms which represent success states.

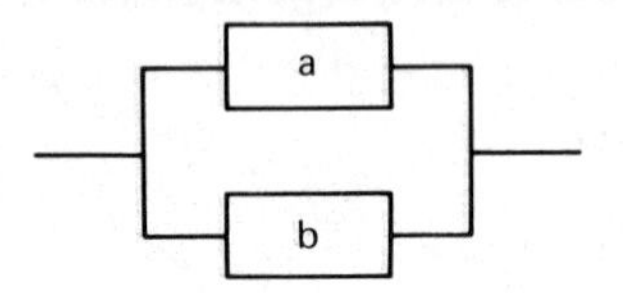

Figure 8.1 Reliability block diagram of two units in parallel reliability

If a and b are constant failure rate units, then

$$R_a(t) = \exp(-\lambda_a t)$$

$$R_b(t) = \exp(-\lambda_b t)$$

and

$$R_s(t) = \exp(-\lambda_a t) + \exp(-\lambda_b t) - \exp[-(\lambda_a + \lambda_b)t]$$

This expression cannot be put in the form $\exp(-\lambda_s t)$ and so the system is not a constant failure rate system, even though it is made up of constant failure rate units.

As an illustration of the advantage of active redundancy, consider that $R_a(t_M) = 0.9$, $R_b(t_M) = 0.9$, then

$$R_s(t_M) = 0.9 + 0.9 - 0.81 = 0.99$$

For the system of Figure 8.1, the MTTF (θ_s) may be found by integration of $R_s(t)$, so that

$$\theta_s = \int_0^\infty R_s(t)\mathrm{d}t = 1/\lambda_a + 1/\lambda_b - 1/(\lambda_a + \lambda_b)$$

while also for the same system:

$$f_s(t) = -\mathrm{d}R_s/\mathrm{d}t = \lambda_a \exp(-\lambda_a t) + \lambda_b \exp(-\lambda_b t) - (\lambda_a + \lambda_b)\exp[-(\lambda_a + \lambda_b)t].$$

For the simple case where $\lambda_a = \lambda_b = \lambda$:

$$R_s(t) = 2\exp(-\lambda t) - \exp(-2\lambda t)$$

$$f_s(t) = 2\lambda(\exp(-\lambda t) - \exp(-2\lambda t))$$

$$\lambda_s(t) = 2\lambda(1 - \exp(-\lambda t))/[2 - \exp(-\lambda t)]$$

and

$$\theta_s = 2/\lambda - 1/2\lambda = 1.5\ \theta_{unit}$$

where $\theta_{unit} = 1/\lambda$.

The reliability function is no longer exponential and its shape is shown in Figure 8.2. For comparison purposes, the $R_s(t)$ has been plotted (broken line) for a series system having the same MTTF. (Hence the areas under the two curves are the same.)

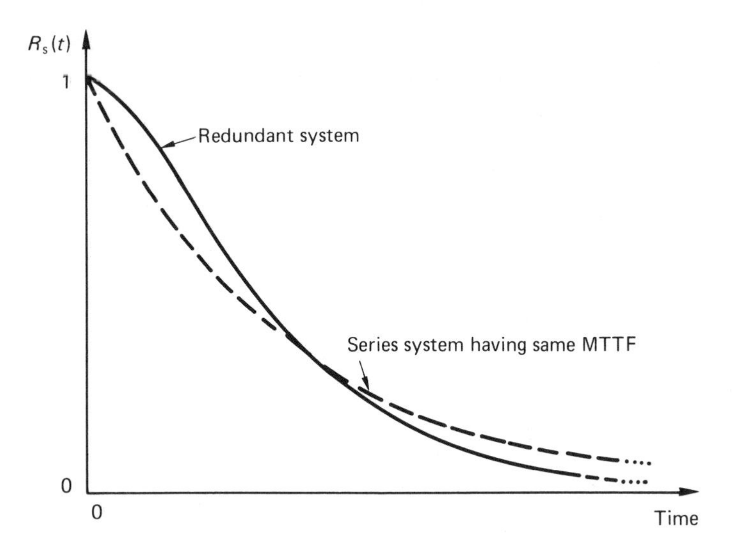

Figure 8.2 Reliability function of a redundant system consisting of two identical units in parallel

It should be noted that in this simple analysis the failure rates have been considered to be invariant regardless of whether a unit is carrying a half or a full load, so that it may reasonably be assumed that the failures of the two units are independent. In practice, the failure rate of a unit may depend upon the load which it is carrying; the failures would not then be independent because failure of one unit would cause the failure rate of the second unit to rise. Analysis of this type of situation is probably best done using Markov analysis.

In the general case where m out of n identical units must stay 'up' for the system to be operational, then the system reliability may be found by taking those terms which refer to the 'up' state from the expansion of $[R(t) + F(t)]^n$. It is conventional to refer to such redundancy configurations as 'm-out-of-n: Good', with the meaning that the system is operational so long as at least m of the parallel units are working. Then

$$R_s = R^n + {}_nC_{n-1}R^{n-1}F + {}_nC_{n-2}R^{n-2}F^2 + \ldots + {}_nC_mR^mF^{n-m}$$

and

$$F_s = {}_nC_{m-1}R^{m-1}F^{n-m+1} + \ldots + {}_nC_1RF^{n-1} + F^n$$

where ${}_nC_r = n!/[(n-r)!r!]$ is the number of combinations of r items taken from n items.

An alternative expression is 'p-out-of-n: Failed', which means that if p or more parallel units fail then the system has failed. Figure 8.3 shows the RBD for a system which is operational so long as at least 2 of the parallel units are operational.

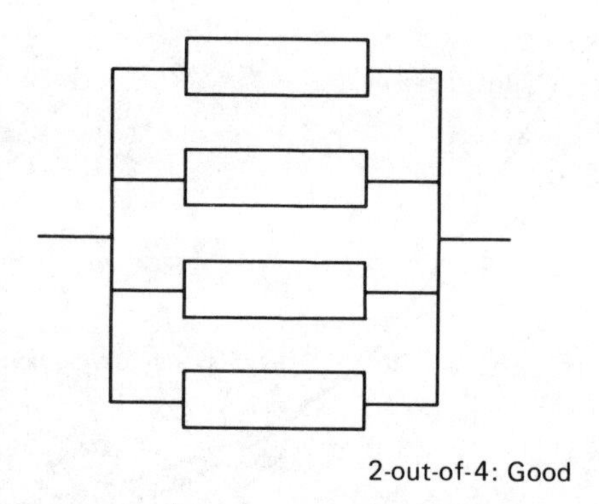

Figure 8.3 An example of m-out-of-n: Good, parallel redundancy

8.3 Active Redundancy: Series-Parallel Reliability

Series-parallel reliability systems may be designed, a simple example of which is shown in Figure 8.4.

The reliability of such a system can be calculated by reducing it to a series system. Thus b and c in parallel reduce to a single unit d for which

$$R_d(t) = R_b(t) + R_c(t) - R_b(t)R_c(t)$$

The system is now represented by a and d in series (Figure 8.5).

The system reliability becomes

$$R_s(t) = R_a(t)\,R_d(t) = R_a(t)\,[R_b(t) + R_c(t) - R_b(t)\,R_c(t)]$$

It should be noted that the unit d does not have a CFR and so it would be incorrect to assume a CFR for the system of Figure 8.4. (The combination of two repairable units in parallel may have an approximate CFR: see section 11.10.)

By a series of simplifications, more complex series-parallel designs may be reduced to a series form.

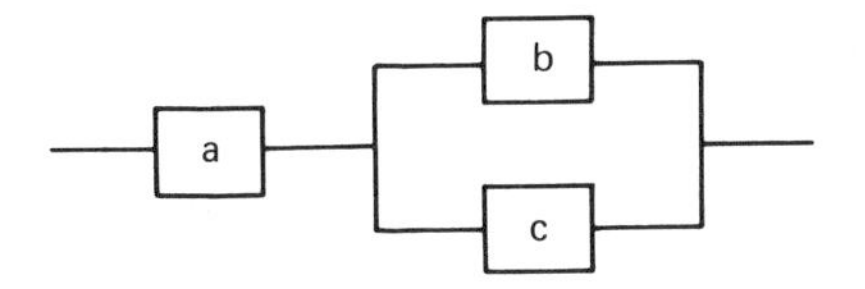

Figure 8.4 RBD of a simple series–parallel system

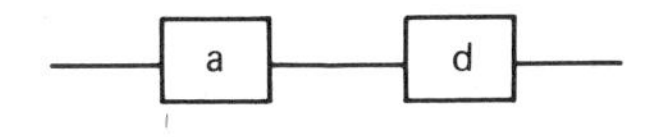

Figure 8.5 RBD of a series system equivalent to that of Figure 8.2

8.4 Passive (Standby) Redundancy

In a passive redundant system, the main unit (a) normally carries the full load and the standby unit (b) is brought into use only when the main unit fails. The simplest type of passive redundant system is illustrated in Figure 8.6.

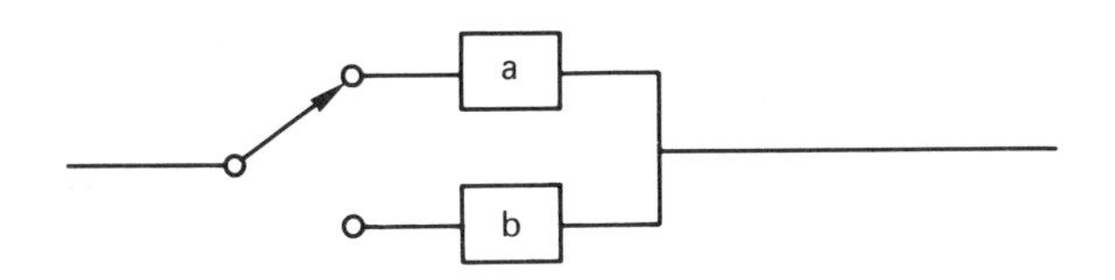

Figure 8.6 Passive redundant system

The most straightforward way of analysing this system is to consider that the system time-to-failure is a random variable which consists of the sum of two random variables: time-to-failure of a and time-to-failure of b. (The possible failure of the switch has been neglected for simplicity.) If we denote the mathematical process of convolution by * then

$$f_s(t) = f_a(t) * f_b(t)$$

from which the other required system functions may be obtained.

If $R_a(t) = R_b(t) = \exp(-\lambda t)$

then it may be shown that

$$R_s(t) = (1 + \lambda t) \exp(-\lambda t)$$

and

$$\theta_s = 2/\lambda = 2\theta_{unit}$$

which is to be expected.

The terms of $R_s(t)$, namely $\exp(-\lambda t)$ and $\lambda t \exp(-\lambda t)$, form the first two terms of the Poisson series with parameter λt, and the further terms of the series

$$\frac{(\lambda t)^2 \exp(-\lambda t)}{2!}, \quad \frac{(\lambda t)^3 \exp(-\lambda t)}{3!} \ldots$$

give the additional terms which must be added to $R_s(t)$ when there are 3, 4 . . . identical constant-failure-rate units in passive (standby) redundancy. This general result may be shown using Laplace transform theory.

The presence of a spare unit may be shown as in Figure 8.7 which represents the RBD of a car.

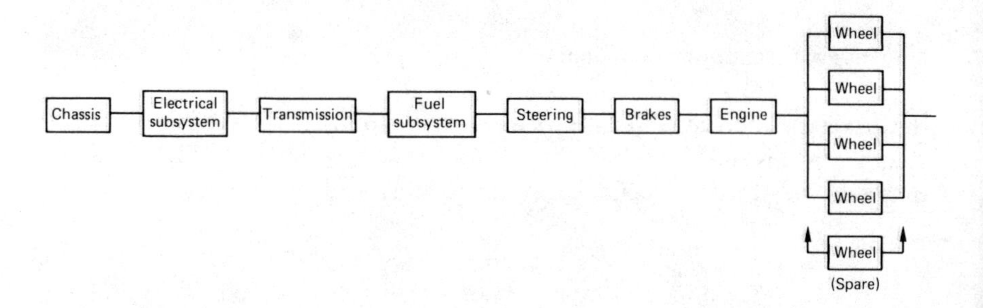

Figure 8.7 RBD of a 4-wheel car carrying one spare wheel

8.5 Combined Active and Passive Redundancy

It is possible to design systems having both active and passive redundancy, as shown in Figure 8.8. Such systems may be analysed by reduction to an equivalent series system.

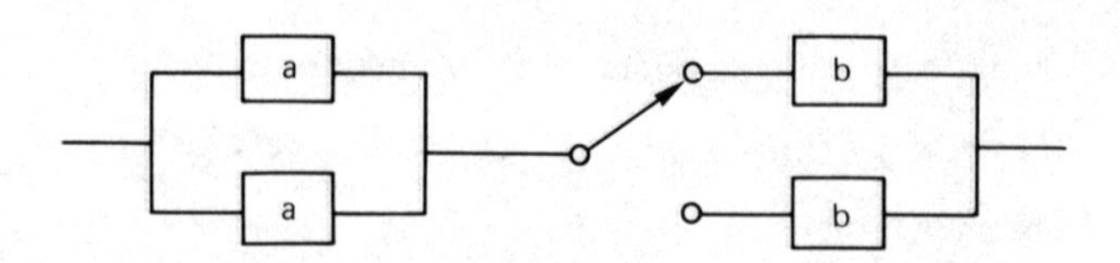

Figure 8.8 A system having both active and passive redundancy

Exercises

8.1. A piece of equipment is cooled by four fans which are kept running continuously. At least two fans are needed to keep the equipment adequately cooled. The probability that a fan will survive for 10 000 hours is 0.9. Given that the probabilities of failure are independent, what is the probability that the equipment will be adequately cooled over a mission time of 10 000 hours?

8.2. You are offered the choice between the two systems whose reliability block diagrams are shown in Figure 8.9. Both systems use constant failure rate components. System (A) uses parallel redundancy; probabilities of failure of the parallel units are independent and one unit can take the whole load without its failure rate increasing.

Which system would you take if your criterion were to choose (a) highest MTTF; (b) highest reliability over a mission time of 1000 hours?

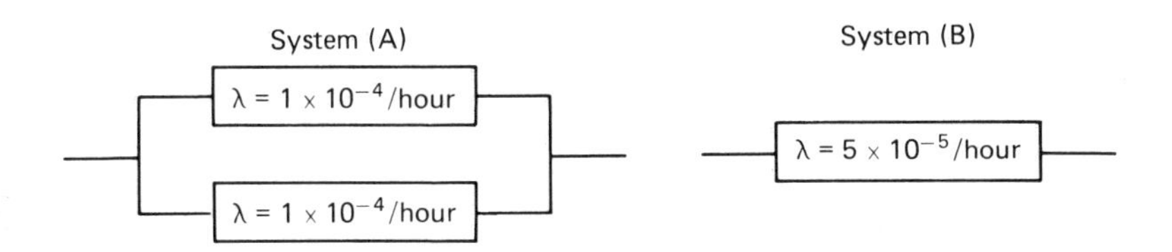

Figure 8.9 RBDs of two competing systems

8.3. Make a rough plot of the expression

$$\lambda_s(t) = 2\lambda(1 - \exp(-\lambda t))/[2 - \exp(-\lambda t)]$$

using (say) $\lambda = 1/10\,000$/hour for the failure rate of two identical units in parallel redundancy. Explain the results for $\lambda_s(0)$ and $\lambda_s(\infty)$.

9 Fault-tolerant Systems – II

9.1 Introduction

In Chapter 8, systems were analysed which could be reduced to an equivalent series-parallel form. However there are systems which cannot be reduced in this way. An example of such a system is shown in Figure 9.1 in which the arrows indicate how the system works; it is seen that the system is 'up' so long as either (a and c) or (b and d) or (b and c) are 'up'.

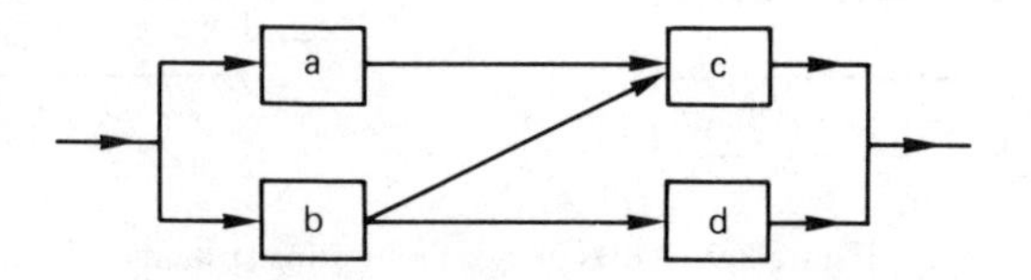

Figure 9.1 A non-series–parallel system

This system cannot be reduced to the form shown in Figure 9.2 because this would imply that the system would be 'up' when only a and d are 'up', and this does not agree with the meaning of the original system diagram of Figure 9.1.

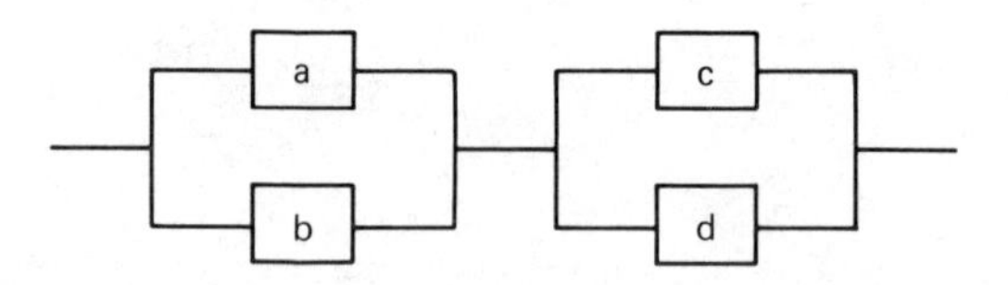

Figure 9.2 A series–parallel system which is not equivalent to that of Figure 9.1

9.2 Use of the Binomial Expansion

An analysis of Figure 9.1 can be carried out using the system states but this is somewhat laborious. Writing a for R_a and $\overline{a}$ for F_a etc., we find that the possible system states are given by the binomial expansion of $(a + \bar{a})(b + \bar{b})(c + \bar{c})(d + \bar{d})$. Of these 16 terms, 8 correspond to the system being up, namely, $abcd$, $\bar{a}bcd$, $a\bar{b}cd$, $ab\bar{c}d$, $abc\bar{d}$, $\bar{a}b\bar{c}d$, $\bar{a}bc\bar{d}$ and $a\bar{b}c\bar{d}$. Thus we can put R_s equal to the sum of these terms: it is convenient to re-arrange them as follows:

$$R_s = (abcd + abc\bar{d}) + ab\bar{c}d + \bar{a}bcd + (a\bar{b}cd + a\bar{b}c\bar{d}) + \bar{a}b\bar{c}d + \bar{a}bc\bar{d}$$

Now $abcd + abc\bar{d} = abc(d + \bar{d}) = abc$. Using this and similar simplifications gives:

$$\begin{aligned} R_s &= abc + ab\bar{c}d + \bar{a}bcd + a\bar{b}c + \bar{a}b\bar{c}d + \bar{a}bc\bar{d} \\ &= (abc + a\bar{b}c) + (\bar{a}bcd + \bar{a}bc\bar{d}) + (\bar{a}b\bar{c}d + ab\bar{c}d) \\ &= ac + \bar{a}bc + b\bar{c}d \end{aligned}$$

Reverting now to the use of R_a, R_b etc., we can put

$$\begin{aligned} R_s &= R_a R_c + R_b R_c (1 - R_a) + R_b R_d (1 - R_c) \\ &= R_a R_c (1 - R_b) + R_b (R_c + R_d - R_c R_d) \end{aligned}$$

9.3 Use of a Probability Map

A probability map (for a very good exposition of probability maps, see Hurley (1963)) is similar to a Karnaugh map in layout. In the probability map, each cell represents the probability of a particular system state. The system of Figure 9.1 has 16 possible states and so the probability map for this system has 16 cells as shown in Figure 9.3. (Standard forms of probability maps are given in Hill and Peterson (1981).)

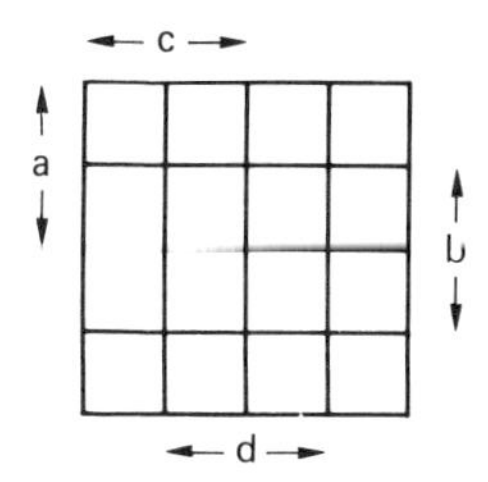

Figure 9.3 A probability map

If a row or column of the map is labelled with a letter, then each cell of that row or column contains that letter; if not, each cell contains the negation of that letter. Thus the top left-hand cell of Figure 9.3 can be labelled $\bar{a}\bar{b}\bar{c}\bar{d}$. Owing to the way in which the map is constructed each cell represents a different system state and the sum of the probabilities of all cells is unity.

It has been shown previously that eight of the possible system states are 'up' states and these have been marked with a '1' in Figure 9.4.

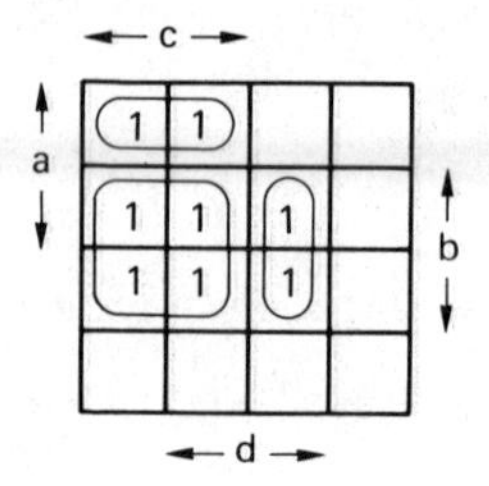

Figure 9.4 The probability map of Figure 9.3 with labelled 'up' states

Any two adjacent cells in a probability map may be grouped together to produce a term with one fewer member than was contained in a single cell. Thus the two top marked cells in Figure 9.4 may be grouped together:

$$a\bar{b}c\bar{d} + a\bar{b}cd = a\bar{b}c(\bar{d} + d) = a\bar{b}c$$

Similarly, four cells in a line or a square may be grouped together. The four cells shown grouped in Figure 9.4 produce

$$abc\bar{d} + abcd + \bar{a}bc\bar{d} + \bar{a}bcd$$

$$= abc + \bar{a}bc$$

$$= bc$$

The remaining two vertically adjacent cells produce:

$$ab\bar{c}d + \bar{a}b\bar{c}d = b\bar{c}d$$

The system reliability is found by summing the grouped terms to be:

$$a\bar{b}c + bc + b\bar{c}d$$

If a is replaced by R_a etc., then we obtain:

$$R_s = R_aR_c(1 - R_b) + R_bR_c + R_bR_d(1 - R_c)$$

$$= R_aR_c(1 - R_b) + R_b(R_c + R_d - R_cR_d)$$

It should be noted that although overlapping of grouped terms is permissible in a Karnaugh map, it is not permissible in a probability map since this would entail counting some system state probabilities twice over. This is why R_s does not equal $R_aR_c + R_bR_c + R_bR_d$.

Probability maps are not, in general, very suitable for the analysis of complex systems since the maps become large and unwieldy. The method which is presented here is very useful as a teaching tool since it presents an easily-understood explanation of how and why the cells in a probability map can be grouped to give a reliability expression.

9.4 Checks on Reliability Expressions

Two checks can be made in general for the validity of an algebraic expression of reliability:

(a) If all the R terms are set equal to zero (equivalent to every unit being in a failed state), then the resultant expression should be zero. Thus we would not expect the expression $0.5 + R_a + R_b R_c$ to be a valid reliability expression, since there appears to be a probability of system success when all the units have failed.

(b) If all the R terms are set to unity (all the units are operational), then we should expect the system reliability to be unity as well. If we obtained the expression $R_a + R_b - 2R_a R_b$ for system reliability, we should be quite sure that there was an error since putting $R_a = R_b = 1$ yields zero probability of system success!

9.5 Bennetts's Method

In Bennetts's original paper (see Bennetts (1975)) he starts from analysis of a fault tree. For simplicity we shall illustrate his method starting from the minimal path sets for a system.

By definition:

> A path set is a group of branches which form a connection between input and output when traversed in the arrowed direction

If no node is traversed more than once in tracing out a path set, that path set is minimal.

We start with the complete set of minimal path sets, which for the system shown in Figure 9.1 is $BD + BC + AC$. We cannot write the system reliability as $bd + bc + ac$ because this would entail counting some of the cells of the probability map (figure 9.3) twice over.

One possible (although rather tedious) course would be to take the canonical expansion

$$BD(A + \bar{A})(C + \bar{C}) + BC(A + \bar{A})(D + \bar{D}) + AC(B + \bar{B})(D + \bar{D})$$

and eliminate all duplicated terms.

Bennetts adopts an alternative approach of making the terms of the complete set of minimal path sets disjoint; the final expression can then be turned into a reliability expression because there is no overlapping. The terms are made disjoint in pairs: first and second, first and third, first and fourth . . . etc., second and third, second and fourth . . . etc. . . . $(n-1)$th and nth.

The method of making two terms disjoint is as follows:

If first path contains one branch (A) which is not in the second path, then multiply second path by $\bar{A}$.
If first path contains two branches (AB) which are not in the second path, then multiply second path by $(\bar{A}+A\bar{B})$
If first path contains three branches (ABC) which are not in the second path, then multiply second path by $(\bar{A}+A\bar{B}+AB\bar{C})$
If first path contains four branches $(ABCD)$ which are not in the second path, then multiply second path by $(\bar{A}+A\bar{B}+AB\bar{C}+ABC\bar{D})$
etc.

In making the terms disjoint, some terms may be generated which can be absorbed by a preceding term (for example, $BC\bar{D}$ absorbs $ACB\bar{D}$ – see probability map); the possibility of this happening should be explored at the end of each 'pass'.

Finally, when all terms are pairwise disjoint the final expression can be converted into a reliability function.

Consider the system shown in Figure 9.1. The complete set of minimal path sets is $BD+BC+AC$.

First pass: make BD and all subsequent terms disjoint.

Step 1: Make BD and BC disjoint.
B is a common branch but D is not contained in BC, hence multiply BC by $\bar{D}$ giving $BC\bar{D}$.

Step 2: Make BD and AC disjoint.
AC does not contain B or D, hence multiply AC by $(\bar{B}+B\bar{D})$ giving $AC(\bar{B}+B\bar{D})$.

End of First pass: $BD+BC\bar{D}+AC(\bar{B}+B\bar{D})=BD+BC\bar{D}+AC\bar{B}+ACB\bar{D}$.

Check for absorption: $ACB\bar{D}$ is contained in $BC\bar{D}$ and can be dropped, yielding $BD+BC\bar{D}+AC\bar{B}$.

Second pass: Make $BC\bar{D}$ and $AC\bar{B}$ disjoint.

Step 1: $BC\bar{D}$ and $AC\bar{B}$ are already disjoint since $BC\bar{D}$ contains B and $AC\bar{B}$ contains $\bar{B}$.

Result: $BD+BC\bar{D}+AC\bar{B}$.

Reliability function is then: $R_bR_d+R_bR_c(1-R_d)+R_aR_c(1-R_b)$

$$=R_aR_c(1-R_b)+R_b(R_c+R_d-R_cR_d)$$

as has been obtained by previous methods.

9.6 Use of the Theorem of Total Probabilities (Factorising Theorem)

The Theorem of Total Probabilities states that:

$$\Pr(S) = \Pr(S \mid E_1)\Pr(E_1) + \Pr(S \mid E_2)\Pr(E_2) + \ldots + \Pr(S \mid E_n)\Pr(E_n)$$

where $\Pr(S)$ is the probability of an event S,

$\Pr(S \mid E_i)$ is the conditional probability of event S given that event E_i has occurred.

$\Pr(E_i)$ is the probability of an event E_i and it is necessary that the events $E_1, E_2 \ldots E_n$ are mutually exclusive (only one can happen) and exhaustive (one or other must happen) so that

$$\Pr(E_1) + \Pr(E_2) + \ldots \Pr(E_n) = 1$$

Reverting to Figure 9.1, let us choose $n = 2$ and

event E_1 = unit b 'up'

event E_2 = unit b 'down'

(events E_1 and E_2 are mutually exclusive and exhaustive).

We now equate $\Pr(S)$ with the reliability of the system (which is a probability) so that

$$R_s = \Pr(S \mid b)\, R_b + \Pr(S \mid \bar{b})\, F_b$$

$\Pr(S \mid b)$ is the probability that the system is 'up' given that b is up. (We can call b the 'pivotal' unit.) It is thus the reliability of the subsystem shown in Figure 9.5 which has been derived from Figure 9.1 by replacing unit b by a short-circuit.

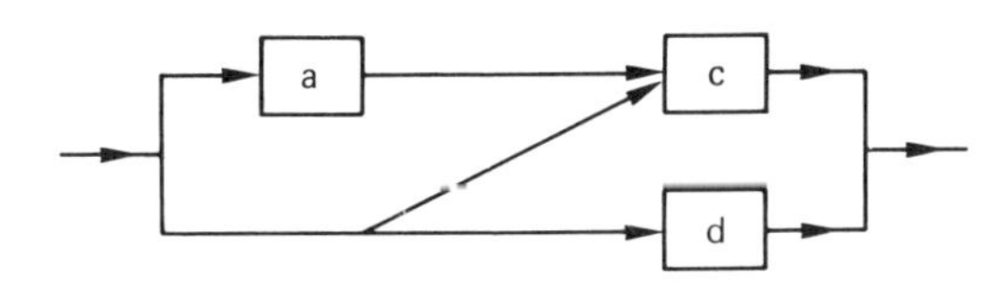

Figure 9.5 Figure used for the evaluation of $\Pr(S \mid b)$

It is seen that unit a has no effect on this subsystem, the reliability of which is simply that of units c and d in parallel; that is

$$\Pr(S \mid b) = R_c + R_d - R_c R_d$$

$\Pr(S \mid \bar{b})$ is derived in a similar way except that the unit b is now replaced by an open-circuit to give the subsystem of Figure 9.6 from which it can be seen that

$$\Pr(S \mid \bar{b}) = R_a R_c$$

Reverting to the expression for system reliability we obtain

$$R_s = (R_c + R_d - R_c R_d) R_b + R_a R_c (1 - R_b)$$

which agrees with the expression obtained by the more lengthy binomial expansion.

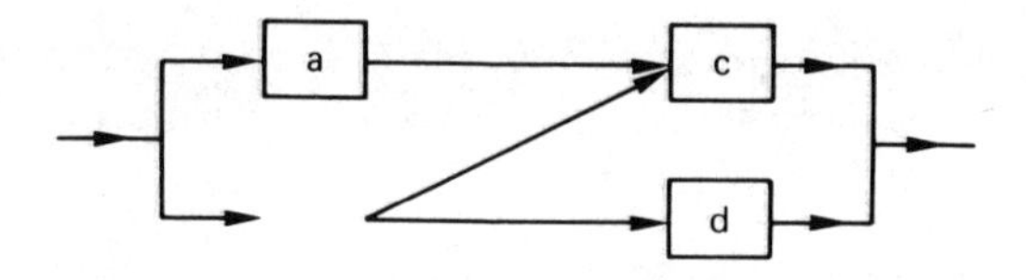

Figure 9.6 Figure used for the evaluation of $\Pr(S \mid \bar{b})$

If we had chosen a unit other than b (the pivotal unit) to give our set of exhaustive and mutually exclusive events, we should have obtained the same answer. However, choice of units a or d as pivotal unit involves a second use of conditional probabilities; choice of c yields a simplification similar to that above.

In general, it appears best to choose as pivotal unit one which has a multiplicity of inputs or outputs.

9.7 Note on the Calculation of Unreliability

The methods quoted in Chapters 8 and 9 have been concerned with the calculation of the reliability function. In the analyses of some systems it may be easier to calculate the unreliability rather than the reliability; the above methods can be used for this purpose when suitable modifications are made.

Exercises

9.1. Find the reliability of the system having the reliability block diagram shown in Figure 9.7 in terms of the reliabilities of the units a, b, c, d, e, using any of the methods described in this chapter.

9.2. Analyse the system of Figure 9.1 using the Theorem of Total Probabilities (as in section 9.6), but using unit c as the pivotal unit.

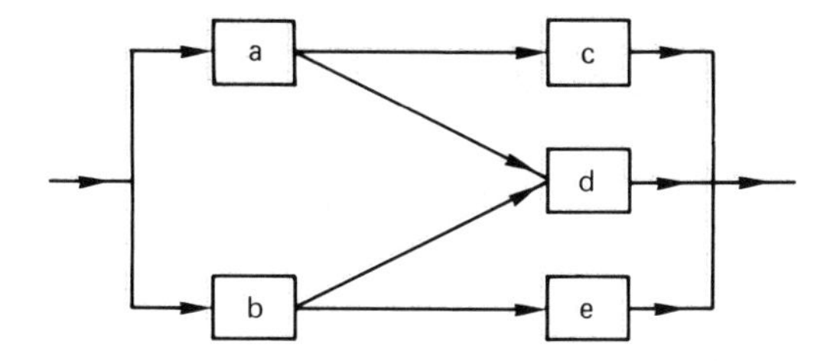

Figure 9.7 Reliability block diagram for use in Exercise **9.1**

10 The Behaviour of Maintained Systems

10.1 Maintenance

It is customary to consider two types of maintenance: corrective and preventive maintenance. Corrective maintenance is concerned with repair actions after a failure has occurred (although if the system contains redundancy the system may not have failed). Preventive maintenance is that concerned with preventing failures (before they occur) by replacement before wear-out, regular oiling etc. Normal maintenance activities will usually consist of both types of maintenance work.

In this chapter, basic ideas on maintenance are given as an introduction to Markov analysis and spare parts provisioning in the next chapters. We return to the more general study of maintenance in Chapter 14 ('Maintainability').

10.2 Corrective Maintenance (Restoration)

10.2.1 Down Time and Restoration Time

It is customary to refer to a system as being 'up' if it is working and 'down' if it has failed. By extension of the usage, 'up time' and 'down time' are used to describe periods of time during which a system is operational or not. A complex system with many failure modes will have (over a period of time) a set of up and down times for each failure mode.

Consider a system which has a repair centre staffed by repairmen who are responsible for repair work in a number of remote sites. An equipment failure causes an alarm to be raised at the repair centre. There may be some time delay (the 'response delay time') before a repairman departs for the site, owing to all repairmen being occupied when the alarm is raised; the travelling time taken to reach the site adds a further delay. The repairman then takes time to repair the equipment (the 'active repair time'). The equipment down time thus comprises three separate times:

Response delay time (the mean of these times being the mean response delay time MRDT)
Travelling time (mean MTT)
Active repair time (mean MART)

Each of these times is a random variable so that the down time, which is given by

down time = response delay time + travelling time + active repair time

is also a random variable. There is a general result in the theory of statistics which states that when random variables are summed, the mean of the sum is equal to the sum of the mean irrespective of the forms of the distributions. Thus the mean down time (MDT) is given by

MDT = MRDT + MTT + MART

Some authors use the term MTTR (mean time to repair), but its use can cause confusion (it can be the MDT or the MART) and its use is not recommended.

It is usual to define a post-repair administrative time (PAT) which is the time taken after the repair is finished for the repairman to write up his report and possibly make other checks (such as for dormant failures) before leaving the site. The down time does not include the PAT.

In what follows, the term 'restoration time' is taken to be synonymous with 'down time'.

10.2.2 Restoration Functions

It is possible to define the distribution of time to restore, the restoration density function and the restoration rate in a manner closely analogous to $F(t)$, $f(t)$ and $\lambda(t)$ for failure. Use of the term 'restoration rate' then avoids the necessity for reference to 'down rate', the meaning of which would be obscure and possibly ambiguous.

The corresponding failure and restoration functions are shown in Figure 10.1 where an exponential distribution has been assumed in both cases, the failure rate being λ and the restoration rate μ. Use of a constant restoration rate may not strictly conform to what is found in practice, but its use is so helpful (particularly in Markov analysis) and the errors involved (if $\mu \gg \lambda$ which is normally the case) so small, that this assumption is often made.

It should be noted that although the *shapes* of the two sets of graphs are identical, the *scales* will in general be vastly different. For example, λ might be of the order of 1/10 000 hours and μ of the order of 1/10 hours.

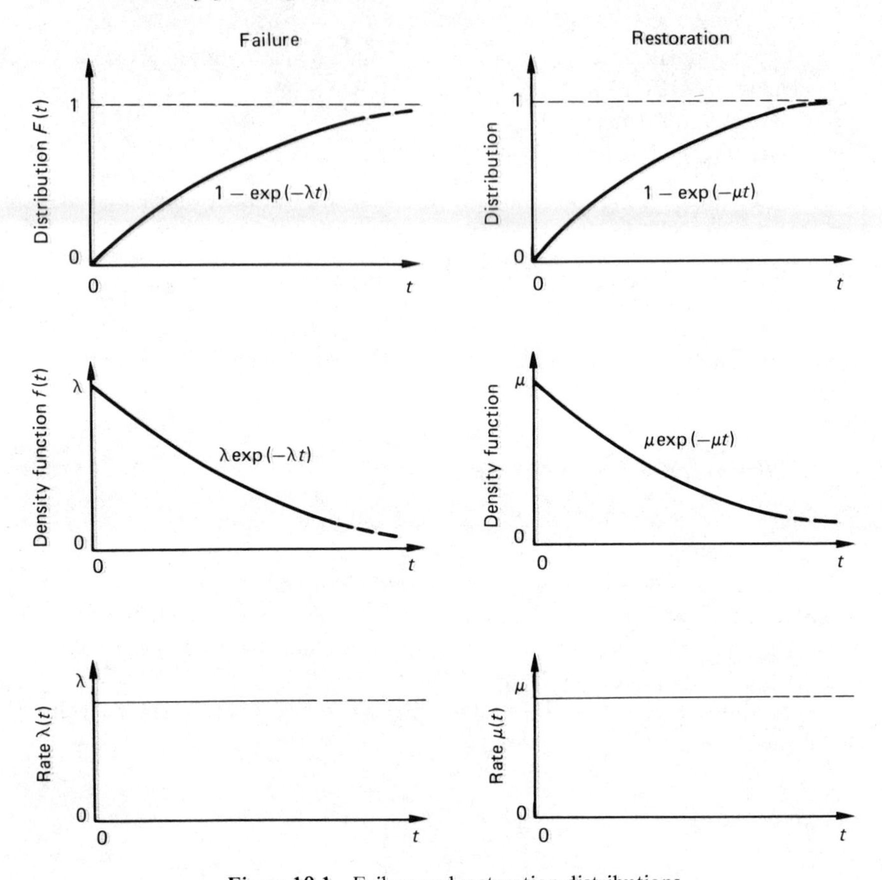

Figure 10.1 Failure and restoration distributions

10.2.3 System Parameters: MUT, MDT, MCT

A maintained system is more complex than a simple system for which the time to failure is often the important parameter. Consequently, it is necessary to be careful over definitions of the various parameters; since MTBF can cause a good deal of confusion, its use is not recommended. It is recommended that mean cycle time (MCT) be used, although which cycle is being considered must also be stated together with the failure mode to which it refers. As a simple example of the use of the term 'cycle time', consider a system that has states '0' and '1' as 'up' states and state '2' as a 'down' state. The state of the system as a function of time is shown in Figure 10.2. Maintained systems generally achieve a 'steady-state' condition some time after switch-on (see Chapter 11) and it is assumed in the figure that the steady-state condition has been reached before the log of system states is started. From the figure, it is seen that successive cycle times $CT_1 = UT_1 + DT_1$, $CT_2 = UT_2 + DT_2$, etc. The mean of CT_1, CT_2 ... is the MCT.

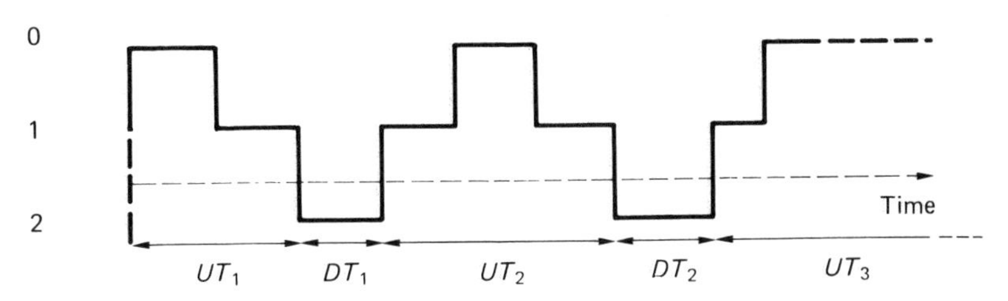

Figure 10.2 Plot of system 'up' and 'down times' against time

10.2.4 Availability and Maintainability

If a log were kept of a large number of systems (see Figure 11.5), it would be possible to plot the proportion of systems which were working ('up') as a function of time. In a manner analogous to the use of the Reliability Function, it is thus possible to define an Availability Function $A(t)$ for systems and to use this for predicting system behaviour. For maintained systems, its slope is no longer necessarily non-positive since failed systems may be restored and brought back into use. The formal definition of $A(t)$ is:

(Instantaneous) Availability, $A(t)$ = Probability that a system will be able to perform its required function at a stated instant in time.

Unavailability $U(t)$, as a function of time, is the complement of $A(t)$:

(Instantaneous) Unavailability, $U(t) = 1 - A(t)$.

The functions $A(t)$ and $U(t)$ are most easily calculated using Markov analysis (see next chapter). Since all systems are assumed to be up at time $t = 0$, it follows that $A(0) = 1$ and $U(0) = 0$. As time goes on, systems will start to fail and $A(t)$ will fall as shown in Figure 11.4. After some time, the steady-state condition will be achieved. This 'Steady-state' Availability is defined by:

$$\text{Steady-state Availability } A_{ss} = \lim_{t \to \infty} [A(t)]$$

and correspondingly

$$\text{Steady-state Unavailability } U_{ss} = \lim_{t \to \infty} [U(t)]$$

$$= 1 - A_{ss}$$

If we define the steady-state availability of a single system as MUT/(MUT + MDT), then it can be shown that this is also A_{ss} as defined above for a group (a 'population' in statistical terms) of systems (see also section 11.5). Thus

$$A_{ss} = \text{MUT}/(\text{MUT} + \text{MDT})$$

and in an equivalent manner

$$U_{ss} = \text{MDT}/(\text{MUT} + \text{MDT})$$

10.2.5 Restoration of Series Reliability Systems: Memoryless Property

It should be noted that for a series system of CFR components, the MTTF (or equivalently the MUT) is unaffected by corrective maintainance which affects only the MDT and the MCT.

For a redundant system, however, the MUT may be greatly affected by the maintenance. This can be seen qualitatively by considering a simple system of two units in parallel reliability; the shorter the restoration time of a failed unit, the less likely it is that the other unit will also fail during this time and the longer will be the system MUT. If a series system is constructed entirely of CFR components, then when one component fails it may be replaced by another CFR component while all the other components may be considered to be 'as good as new'. Thus the system as a whole has the same memoryless property as if it were a single CFR component with an equivalent failure rate (see section 3.14).

10.2.6 Repair of Non-series Reliability Systems

So long as non-series systems are constructed of CFR subsystems and the restoration rate can be considered to be constant, then Markov analysis can be used (see the next chapter). If these conditions do not hold, then a special analysis or simulation may be necessary. However, consideration should always be given to whether the necessary conditions are approximately satisfied sufficiently closely for a Markov analysis to be valid.

10.2.7 Distribution of Mean Cycle Time (MCT)

The Mean Cycle Time has been defined as the mean of a series of cycle times $\text{CT}_1, \text{CT}_2 \ldots$ for a single system. The series of cycle times can be considered to be random selections from the cycle time probability density function. The MCT is clearly the mean value of this density function and can be found by convolution (see sections 3.16.2 and 3.16.3) of the failure and restoration density functions. Once it is known, then it can be used for calculation of any desired reliability information (such as expected number of failures in a given time T). It should be noted that even when λ and μ are constant, the Cycle Time is not exponentially distributed. Its pdf is

$$\mu\lambda\,[\exp(-\lambda t) - \exp(-\mu t)]\,/(\mu - \lambda)$$

the mean of which (the MCT) is $1/\lambda + 1/\mu$ (see Exercise **10.1**). This is MUT + MDT, which is what would be expected intuitively.

In most practical cases, restoration times are very much shorter than failure times, that is, $\mu \gg \lambda$; typically, μ may be 0.2/hour (MDT = 5 hours) while λ may be 0.0001/hour (MUT = 10 000 hours). Thus it is usually possible to calculate reliability parameters using the assumption that the MCT (or MTBF) is exponentially distributed with parameter λ. The methods of Chapter 15 can then be used for the calculation of expected number of failures in time T, composite failure pattern etc.

In more complex cases, Markov analysis (see next chapter) or renewal theory can be used. The failure-repair-failure- . . . . cycle is known in renewal theory as an Alternating Renewal Process.

10.3 Preventive Maintenance

10.3.1 Replacement before Wear-Out

In the field of telecommunications, chief interest in preventive maintenance is concerned with how to deal with limited-life components like lamps and electro-mechanical components. The failure density of lamps is generally considered to be approximately normally distributed with a mean life of m and a standard deviation of s. Interest is then generally centred on what time the whole population should be replaced ($m - 3s$, $m - 2s$, say) in order to avoid a sharp increase in maintenance costs corresponding to a rapid fall in the reliability which occurs at about this time. If the maintenance action is carried out early enough, it may be possible to treat the device as if it were a CFR component for reliability predictions.

10.3.2 Regular Maintenance

When all components are instantaneously replaced (or correspondingly made 'as good as new') at times t_R, $2t_R$, $3t_R$ etc., then the system availability curve will repeat itself indefinitely as shown in Figure 10.3.

It is not now possible to define a mean time to failure for the components, since many are replaced before they fail. But it is possible to define a mean up time for the positions in which the components are placed, and this parameter is certainly of interest because it defines the mean time that will elapse between failures at the same position.

If we imagine each position to be equipped with a clock which registers the elapsed time between replacement of a failed component and occurrence of the next failure in that position, then after a sufficiently long period of time we can put

$$\text{positional mean up time (PMUT)} = \frac{\text{sum of clock readings}}{\text{number of failures}}$$

$$= \frac{\text{sum of clock readings/(number of positions} \times \text{number of replacement cycles)}}{\text{number of failures/(number of positions} \times \text{number of replacement cycles)}}$$

$$= \frac{\text{average time registered/replacement cycle}}{\text{average fraction of failures/replacement cycle}}$$

$$= \frac{\int_0^{t_R} R(t)\mathrm{d}t}{1 - R(t_R)}$$

Note that when $R(t) = \exp(-\lambda t)$, then

$$\int_0^{t_R} R(t)\mathrm{d}t = [-\exp(-\lambda t)/\lambda]_0^{t_R} = [1 - \exp(-\lambda t_R)]/\lambda$$

$$= [1 - R(t_R)]/\lambda$$

Then PMUT $= 1/\lambda$.

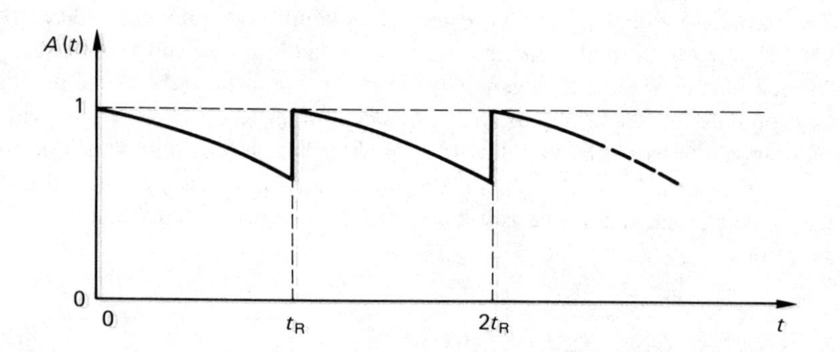

Figure 10.3 Availability of system with regular preventive maintenance

This is the result which we should expect, and illustrates the fact that no advantage is to be gained by replacing a CFR component or system before failure (so long as the component or system is within its useful life).

Regular preventive maintenance is, however, clearly advantageous when limited-life components or systems are used. The maintenance interval t_R should then be chosen to be at a time on the bathtub curve which is before the onset of wear-out.

Exercises

***10.1.** Multiply together the Laplace transforms of the exponential failure density function $\lambda \exp(-\lambda t)$ and the exponential restoration density function $\mu \exp(-\mu t)$. By taking the inverse Laplace transform of this product, find the density function of the cycle time (failure-repair- . . . etc.) and show that the MCT is $1/\lambda + 1/\mu$ (as given in section 10.2.7).

10.2. Calculate the positional mean up time (PMUT), given that

$$R(t) = 1 - t/t_E \quad 0 \leqslant t \leqslant t_E$$

$$= 0 \quad t > t_E$$

where t_E = 20 000 hours and t_R = 10 000 hours. Compare the value you obtain with the corresponding MTTF (see Exercise **3.4**).

11 Elements of Markov Analysis

11.1 Introduction

Markov analysis can be applied when transitional rates between system states are constant (that is, independent of time). It is usual to draw a Markov (state) diagram which shows the various states which a system may enter as a result of failure or restoration of the units comprising the system together with the failure (or restoration) rates associated with the transitions.

11.2 Simple Case: No Restoration

The simplest case to consider is that shown in Figure 11.1 where the system consists of a single unit a with a constant failure rate λ_a. The two possible system states are designated 0 and 1. The 'up' (working) state is indicated by being shown in a circle while the 'down' (failed) state is indicated by being shown in a square (or sometimes a rectangle).

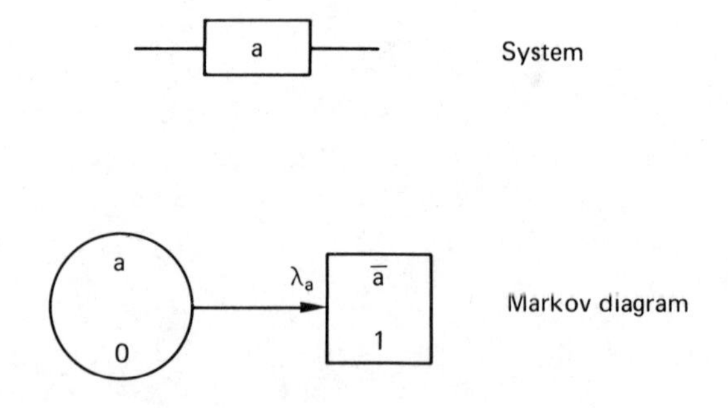

Figure 11.1 A simple system and corresponding Markov diagram

We shall use $P_0(t)$ and $P_1(t)$ to denote the probabilities that the system is in states 0 and 1 respectively. Both of these probabilities are functions of time and we shall take it as axiomatic that $P_0(0) = 1$.

Now since the failure rate λ_a is constant, the probability of failure from state 0 to state 1 (given that the system is in state 0 at time t) is independent of the past history of the system and for a small interval of time Δt this probability is approximately $\lambda_a \Delta t$ (see definition of failure rate).

The probability that the system is in state 1 at a time $t + \Delta t$ can now be expressed as the sum of two separate probabilities: that the system was in state 1 at time t and stayed there (since there was nowhere else to go); and that the system was in state 0 at time t and passed into state 1 during the interval t to $(t + \Delta t)$. These two probabilities are $P_1(t)$ and $P_0(t)\,\lambda_a \Delta t$.

Thus we can put

$$P_1(t + \Delta t) \approx P_1(t) + P_0(t)\,\lambda_a \Delta t \qquad (11.1)$$

Re-arrangement gives

$$\frac{P_1(t + \Delta t) - P_1(t)}{\Delta t} \approx P_0(t)\,\lambda_a$$

and taking the limit as $t \to 0$ gives

$$\frac{\mathrm{d}P_1}{\mathrm{d}t} = \lambda_a P_0(t) \qquad (11.2)$$

where the equality sign can now replace the 'approximately equals' sign.

We can make use of another property of the system: it must be either in state 0 or state 1 at all times, so that for any value of t:

$$P_0(t) + P_1(t) = 1 \qquad (11.3)$$

Differentiating with respect to t and re-arranging gives

$$\frac{\mathrm{d}P_0}{\mathrm{d}t} = -\frac{\mathrm{d}P_1}{\mathrm{d}t}$$

and when $\mathrm{d}P_1/\mathrm{d}t$ is eliminated between this and equation (11.2) we obtain

$$\frac{\mathrm{d}P_0}{\mathrm{d}t} = -\lambda_a P_0(t)$$

the solution of which (using $P_0(0) = 1$) is

$$P_0(t) = \exp(-\lambda_a t)$$

as we should expect. (Alternatively, this equation could be obtained from: $P_0(t + \Delta t) \approx P_0(t)[1 - \lambda_a \Delta t]$ which is obtained by considering the probability that the system was in state 0 at time t and did not fail in the interval t to $(t + \Delta t)$.)

Making use of equation (11.2) gives the expected result:

$$P_1(t) = 1 - \exp(-\lambda_a t)$$

11.3 Simple Case: With Restoration

Now consider the system of Figure 11.1 in which restoration is now allowed from the failed state. Let the restoration rate be constant and denoted by μ. (μ is here called the restoration rate rather than repair rate for the reasons given in section 10.2.2.) The Markov diagram becomes that of Figure 11.2.

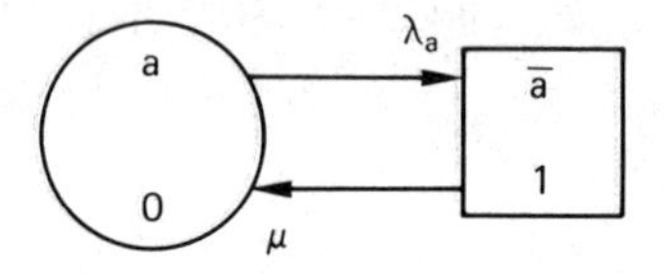

Figure 11.2 Markov diagram of system with restoration

It is usually easier to use double-headed arrows as shown in Figure 11.3. (The use of double-headed arrows considerably simplifies complex Markov diagrams.)

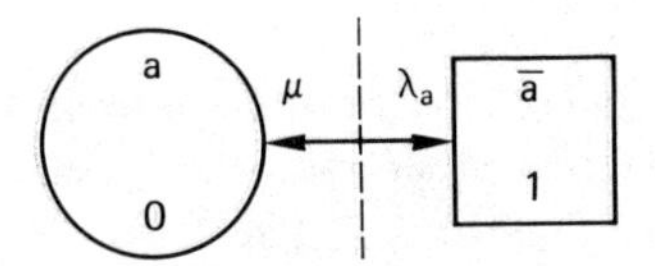

Figure 11.3 Markov diagram with double-headed arrows

Equation (11.1) above must now be modified, since a transition from state 1 to state 0 is now possible. Since the restoration rate is constant, the probability of restoration from state 1 to state 0 (given that the system is in state 1 at time t) is independent of the past history of the system, and for a small interval of time t this probability is $\mu\Delta t$. Correspondingly, the probability of not restoring the system (and not going from state 1 to state 0) must be approximately $1 - \mu\Delta t$.

Thus equation 11.1 must be modified to become:

$$P_1(t + \Delta t) \approx P_1(t)[1 - \mu\Delta t] + P_0(t)\,\lambda_a\Delta t$$

Re-arrangement of this expression gives

$$\frac{P_1(t + \Delta t) - P_1(t)}{\Delta t} \approx -\mu\, P_1(t) + \lambda_a\, P_0(t)$$

and taking the limit as $t \rightarrow 0$ gives

$$\frac{dP_1}{dt} = -\mu\, P_1(t) + \lambda_a\, P_0(t) \qquad (11.4)$$

A similar argument using state 0 instead of state 1 leads to

$$P_0(t + \Delta t) \approx P_0(t)[1 - \lambda_a \Delta t] + P_1(t)\,\mu\Delta t$$

which becomes

$$\frac{dP_0}{dt} = -\lambda_a P_0(t) + \mu P_1(t) \tag{11.5}$$

in the limit.

Equations (11.4) and (11.5) together with equation (11.3), that is

$$P_0(t) + P_1(t) = 1$$

must now be solved.

*11.4 Use of Laplace Transforms

It is simplest to solve equations (11.4), (11.5) and (11.3) using Laplace transforms together with the known system state at $t = 0$, namely:

$$P_0(0) = 1 \quad \text{and} \quad P_1(0) = 0$$

When transformed the three equations become*:

$$s\bar{P}_1(s) = -\bar{P}_1(s) + \lambda_a \bar{P}_0(s)$$

$$s\bar{P}_0(s) - 1 = -\lambda_a \bar{P}_0(s) + \bar{P}_1(s)$$

$$\bar{P}_0(s) + \bar{P}_1(s) = \frac{1}{s}$$

in which $\bar{P}_i(s)$ is used to denote the Laplace transform of $P_i(t)$. These three equations are not independent and only two of them are required to give the solution.

$$\left.\begin{aligned} \bar{P}_0(s) &= \frac{s+\mu}{s(s+\mu+\lambda_a)} = \frac{\mu}{\mu+\lambda_a}\left(\frac{1}{s}\right) + \frac{\lambda_a}{\mu+\lambda_a}\left(\frac{1}{s+\mu+\lambda_a}\right) \\ \bar{P}_1(s) &= \frac{\lambda_a}{s(s+\mu+\lambda_a)} = \frac{\lambda_a}{\mu+\lambda_a}\left(\frac{1}{s}\right) - \frac{\lambda_a}{\mu+\lambda_a}\left(\frac{1}{s+\mu+\lambda_a}\right) \end{aligned}\right\} \tag{11.6}$$

The inverse Laplace transforms of these functions then give $P_0(t)$ and $P_1(t)$. The inverse Laplace transform of $1/s$ is the unit step function $u(t)$. Since we are interested only in the period $0 < t < \infty$, it is sufficient for us to take $u(t) = 1$.

*For consistency in the Laplace transformations, it is necessary to replace equation (11.3) by $P_0(t) + P_1(t) = u(t)$.

11.5 State Probabilities as Functions of Time: Ergodicity

Inversion of equations (11.6) gives

$$P_0(t) = \frac{\mu}{\mu + \lambda_a} + \frac{\lambda_a}{\mu + \lambda_a} \exp[-(\mu + \lambda_a)t]$$

$$P_1(t) = \frac{\lambda_a}{\mu + \lambda_a} (1 - \exp[-(\mu + \lambda_a)t])$$

A partial check on the validity of these equations can be made by noting that:

$$P_0(0) = 1 \quad \text{and} \quad P_0(t) + P_1(t) = 1$$

The variation of $P_0(t)$ and $P_1(t)$ with time is shown in Figure 11.4.

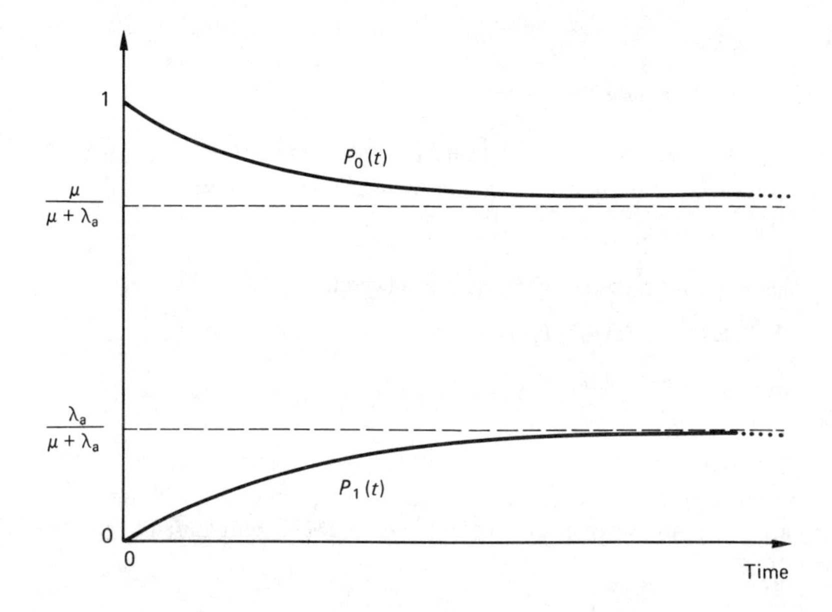

Figure 11.4 Variation of $P_0(t)$ (availability) and $P_1(t)$ (unavailability) with time

The physical 'meaning' of these state probabilities is that if a large number of identical systems were operated simultaneously, all being switched on at time $t = 0$, then $P_i(t)$ represents the proportion of systems in state i at any time t. (In this case $i = 0$ or 1; in the analysis of complex systems, the number of states can be quite large.) The meaning of the state probabilities for $i = 0$ or 1 is illustrated in Figure 11.5.

In those cases where a steady-state is reached (as shown in Figure 11.4), then it is possible to show that the statistics obtained by examining a single system over a long period of time (as indicated in Figure 10.2) are exactly the same as those obtained by examining the instantaneous (steady-state) probabilities. Groups of systems having this property are said to be ergodic (see Allen (1978)).

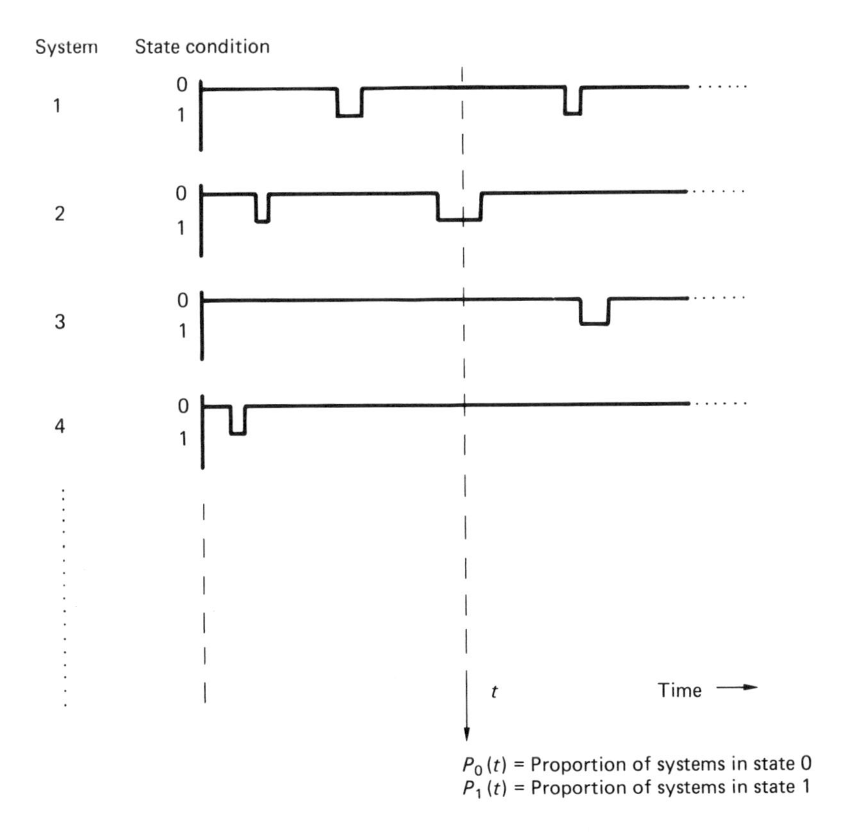

Figure 11.5 Illustration of the meaning of state probability

For the system represented by Figure 11.3 the steady-state condition is represented by

$$\lim_{t\to\infty} P_0(t) = P_0(\infty) = \mu/(\mu + \lambda_a)$$

and

$$\lim_{t\to\infty} P_1(t) = P_1(\infty) = \lambda_a/(\mu + \lambda_a)$$

It is shown in section 11.9 that the MCT is the inverse of the rate of crossing the broken line in Figure 11.3 which (in the steady-state condition) is given by $\lambda_a P_0(\infty)$. Thus the MCT calculated from the state probabilities is $(\mu + \lambda_a)/\lambda_a\mu$

or $1/\lambda_a + 1/\mu$. This is exactly the same as the MCT obtained in section 10.2.7 when the statistics of only one system were considered.

The steady-state availability obtained from the state probabilities is $P_0(\infty)$ (since state 0 is the only 'up' state) and as expected:

$$P_0(\infty) = \mu/(\mu + \lambda_a) = \text{MUT}/(\text{MUT} + \text{MDT})$$
$$= A_{ss}$$

Similarly:

$$P_1(\infty) = \lambda_a/(\mu + \lambda_a) = \text{MDT}/(\text{MUT} + \text{MDT})$$
$$= U_{ss}$$

since state 1 is the only 'down' state.

11.6 Direct Calculation of Steady-state Availability

In a system in which restoration from a failed state is allowed, it is to be expected that a steady-state position will eventually be achieved in which the state probabilities are constant, irrespective of time. Such a condition was shown above in the simple case of Figure 11.3 where the steady-state probabilities are

$$P_0(\infty) = A_{ss} = \frac{\mu}{\mu + \lambda_a}$$

$$P_1(\infty) = U_{ss} = \frac{\lambda_a}{\mu + \lambda_a}$$

When only the steady-state conditions are of interest, equations (11.4) and (11.5) may be simply solved for $P_0(\infty)$ and $P_1(\infty)$ by setting

$$dP_0/dt = 0$$

$$dP_1/dt = 0$$

since when both P_0 and P_1 are constant, their differential coefficients must be zero.

The two differential equations reduce to the same equation:

$$0 = -P_1(\infty) + \lambda_a P_0(\infty)$$

$$0 = -\lambda_a P_0(\infty) + P_1(\infty)$$

These equations can be solved by noting additionally that

$$P_0(\infty) + P_1(\infty) = 1$$

giving

$$P_0(\infty) = A_{ss} = \frac{\mu}{\mu + \lambda_a}$$

$$P_1(\infty) = U_{ss} = \frac{\lambda_a}{\mu + \lambda_a}$$

as was obtained above.

The example quoted here is rather trivial, but the same technique of setting the differential coefficients equal to zero can be used in much more complex cases when solution of a set of differential equations is reduced to solution of a set of linear equations.

In the more complex cases when the states 0, 1, . . ., i are 'up' states and the states $j, k, \ldots, z$ are 'down' states, then

$$A_{ss} = P_0(\infty) + P_1(\infty) + \ldots + P_i(\infty)$$

and

$$U_{ss} = P_j(\infty) + P_k(\infty) + \ldots + P_z(\infty)$$

11.7 The Calculation of MTTFF: Absorbing States

For a system which takes some time to reach equilibrium conditions, the reliability during the (comparatively) early stages can be assessed by calculation of the Mean Time To First Failure (MTTFF), Mean Time To Second Failure (MTTSF) etc.; in the limit, the *differences* between these successive times to failure become constant and equal to the MCT. Of these successive mean times, the MTTFF is usually of most interest, particularly if system failure is a safety hazard. It is the mean of the times that would be obtained by allowing many systems to run from $t = 0$ to the time of first failure. Since it may take some time for a repairable system to reach statistical equilibrium, the MTTFF will not in general equal the MUT. The MTTFF may readily be calculated using Markov analysis by making the failed state (or states) an absorbing state, that is, a state which can be entered but from which departure (by repair) is not allowed.

As a simple example of the calculation, consider the system having the Reliability Block Diagram and Markov diagram shown in Figure 11.6.

In this example, state 3 is an absorbing state. Using the procedures shown above, the following differential equations may be constructed:*

*A useful check on a set of equations such as these is to note that any path represents an 'exit' from one state and also an 'entry' into another state. For this reason, the sums of the coefficients (taken over all the equations) for any state i should be zero. For example, taking $i = 0$ the coefficients of $P_0(t)$ sum to

$$-(\lambda_a + \lambda_b) + (\lambda_a) + (\lambda_b) + (0) = 0$$

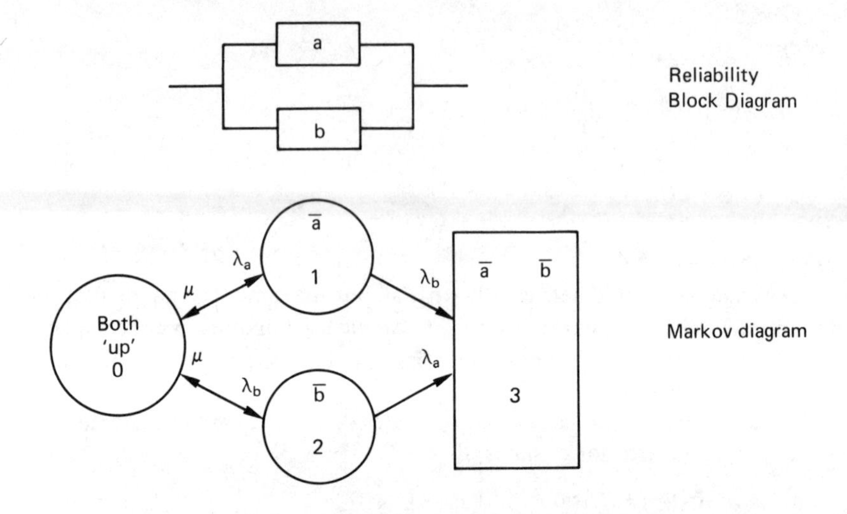

Figure 11.6 Example for calculation of MTTFF

$$\left.\begin{aligned}
\frac{dP_0}{dt} &= -(\lambda_a + \lambda_b)\,P_0(t) + \mu P_1(t) + \mu P_2(t) \\
\frac{dP_1}{dt} &= -(\lambda_b + \mu)\,P_1(t) + \lambda_a\,P_0(t) \\
\frac{dP_2}{dt} &= -(\lambda_a + \mu)\,P_2(t) + \lambda_b\,P_0(t) \\
\frac{dP_3}{dt} &= \lambda_b\,P_1(t) + \lambda_a\,P_2(t)
\end{aligned}\right\} \tag{11.7}$$

Now the reliability* of the system, $R_s(t)$, is the probability that the system is in operation at time t. The states 0, 1 and 2 are the only 'up' states and since the probabilities of being in these states are mutually exclusive, it follows that they can be added to give

$$R_s(t) = P_0(t) + P_1(t) + P_2(t)$$

Hence

$$\text{System MTTFF} = \int_0^\infty R_s(t)\,dt = \int_0^\infty P_0(t)\,dt + \int_0^\infty P_1(t)\,dt + \int_0^\infty P_2(t)\,dt$$

*It is reasonable to consider system reliability since, although the individual units are repairable, the system as a whole (for the purpose of this calculation) is considered to be unrepairable.

If we use the nomenclature

$$\theta_i = \int_0^\infty P_i(t)\,\mathrm{d}t$$

then by integrating both sides of equations (11.7) between 0 and ∞ we obtain

$$P_0(\infty) - P_0(0) = -(\lambda_a + \lambda_b)\,\theta_0 + \mu\theta_1 + \mu\theta_2$$

$$P_1(\infty) - P_1(0) = -(\lambda_b + \mu)\,\theta_1 + \lambda_a\,\theta_0$$

$$P_2(\infty) - P_2(0) = -(\lambda_a + \mu)\,\theta_2 + \lambda_b\,\theta_0$$

$$P_3(\infty) - P_3(0) = \lambda_b\,\theta_1 + \lambda_a\,\theta_2$$

Now at $t = 0$, the system is defined to be in state 0 so that

$$P_1(0) = P_2(0) = P_3(0) = 0$$

At $t = \infty$, the system must be in state 3 since it is an absorbing state. (This follows from making it axiomatic that $R_s(\infty) = 0$.) Hence:

$$P_3(\infty) = 1 \quad \text{and} \quad P_0(\infty) = P_1(\infty) = P_2(\infty) = 0$$

The above equations then become:

$$-1 = -(\lambda_a + \lambda_b)\,\theta_0 + \mu\,\theta_1 + \mu\,\theta_2$$

$$0 = -(\lambda_b + \mu)\,\theta_1 + \lambda_a\,\theta_0$$

$$0 = -(\lambda_a + \mu)\,\theta_2 + \lambda_b\,\theta_0$$

$$1 = \lambda_b\,\theta_1 + \lambda_a\,\theta_2$$

There are four equations in three unknowns (and so the equations are not independent). Taking any three of them yields the solution:

$$\theta_0 = \frac{(\lambda_a + \mu)(\lambda_b + \mu)}{\lambda_a\lambda_b(\lambda_a + \lambda_b + 2\mu)}$$

$$\theta_1 = \frac{\lambda_a}{\lambda_b + \mu}\,\theta_0$$

$$\theta_2 = \frac{\lambda_b}{\lambda_a + \mu}\,\theta_0$$

Hence:

$$\text{MTTFF} = \theta_0 + \theta_1 + \theta_2 = \frac{(\lambda_a + \mu)(\lambda_b + \mu) + \lambda_a(\lambda_a + \mu) + \lambda_b(\lambda_b + \mu)}{\lambda_a\lambda_b(\lambda_a + \lambda_b + 2\mu)}$$

It is often the case that $\mu \gg \lambda_a, \lambda_b$, in which case

$$\text{MTTFF} \approx \frac{[\mu^2 + 2\mu(\lambda_a + \lambda_b)]}{[2\mu\lambda_a\lambda_b]}$$

or $$\text{MTTFF} \approx \frac{\mu}{2\lambda_a\lambda_b}$$

A check on the validity of the complete expression for MTTFF may be obtained by allowing μ to equal zero. This corresponds to the case of no restoration (that is, restoration may be considered to take an infinitely long time) and the MTTFF becomes the MTTF. In this case

$$\text{MTTF} = \frac{\lambda_a^2 + \lambda_a\lambda_b + \lambda_b^2}{\lambda_a\lambda_b(\lambda_a + \lambda_b)} = \frac{1}{\lambda_a} + \frac{1}{\lambda_b} - \frac{1}{\lambda_a + \lambda_b}$$

which is the expected result. (This equality is most easily proved by reducing the right-hand side to a single expression.)

11.8 Calculation of MTTF for a Non-restorable System

A non-restorable system must have at least one absorbing state and so the methods of the previous section can be used.

11.9 Calculation of MUT, MDT and MCT

To illustrate the calculation of MUT, MDT and MCT, the simple series system shown in Figure 11.7 will be used. The two units a and b are given different restoration rates μ_a and μ_b. (It may be necessary to distinguish between the restoration rates if one unit takes appreciably longer to restore than the other.)

The steady-state probabilities for the three states are given by the equations:

$$0 = -\lambda_a P_0(\infty) + \mu_a P_1(\infty)$$

$$0 = -\lambda_b P_0(\infty) + \mu_b P_2(\infty)$$

$$1 = P_0(\infty) + P_1(\infty) + P_2(\infty)$$

which may be solved to give

$$P_0(\infty) = A_{ss} = \frac{1}{1 + \dfrac{\lambda_a}{\mu_a} + \dfrac{\lambda_b}{\mu_b}} \tag{11.8}$$

$$P_1(\infty) = \frac{\dfrac{\lambda_a}{\mu_a}}{1 + \dfrac{\lambda_a}{\mu_a} + \dfrac{\lambda_b}{\mu_b}} \qquad P_2(\infty) = \frac{\dfrac{\lambda_b}{\mu_b}}{1 + \dfrac{\lambda_a}{\mu_a} + \dfrac{\lambda_b}{\mu_b}}$$

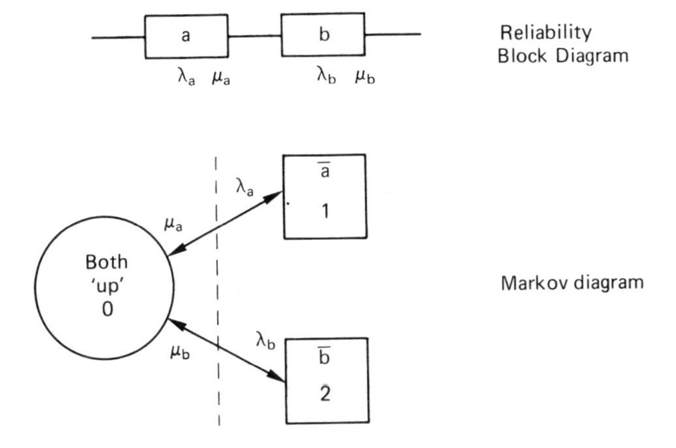

Figure 11.7 Reliability block diagram and Markov diagram for two units in series

so that the steady-state unavailability is

$$U_{ss} = P_1(\infty) + P_2(\infty) = \frac{\dfrac{\lambda_a}{\mu_a} + \dfrac{\lambda_b}{\mu_b}}{1 + \dfrac{\lambda_a}{\mu_a} + \dfrac{\lambda_b}{\mu_b}}$$

The MCT may be found as the reciprocal of the rate at which the broken line in Figure 11.7 is crossed. Under equilibrium conditions, in a long period of time T, state 0 will be occupied for a length of time $P_0(\infty)T$. The mean time between crossings from left to right from state 0 into state 1 is $1/\lambda_a$. Thus the total number of crossings in time T is

$$\frac{P_0(\infty)T}{1/\lambda_a} = \lambda_a P_0(\infty)T$$

and the rate of crossing is

$$\frac{\lambda_a P_0(\infty)T}{T} = \lambda_a P_0(\infty)$$

Similarly, the rate of crossing from state 0 to state 2 is $\lambda_b P_0(\infty)$ so that the total rate of crossing the broken line of Figure 11.7 from left to right is

$$\lambda_a P_0(\infty) + \lambda_b P_0(\infty) = (\lambda_a + \lambda_b) P_0(\infty)$$

Hence under equilibrium conditions, the MCT is the reciprocal of this value and so

$$\text{MCT} = \frac{1}{(\lambda_a + \lambda_b)\, P_0(\infty)} = \frac{1 + \dfrac{\lambda_a}{\mu_a} + \dfrac{\lambda_b}{\mu_b}}{\lambda_a + \lambda_b}$$

(The MCT could equivalently be found as the rate of crossing the broken line of Figure 11.7 from right to left, since under equilibrium conditions, the two rates must be the same.)

It follows that

$$\text{MUT} = A_{ss}\text{MCT}$$

and using equation (11.8) for A_{SS}:

$$\text{MUT} = \frac{1}{\lambda_a + \lambda_b}$$

which is clearly correct. Also

$$\text{MDT} = U_{ss}\text{MCT} = \frac{\dfrac{\lambda_a}{\mu_a} + \dfrac{\lambda_b}{\mu_b}}{\lambda_a + \lambda_b}$$

$$= \left(\frac{\lambda_a}{\lambda_a + \lambda_b}\right)\frac{1}{\mu_a} + \left(\frac{\lambda_b}{\lambda_a + \lambda_b}\right)\frac{1}{\mu_b}$$

This result is reasonable since it represents the weighted average of the two independent 'down' times $1/\mu_a$ and $1/\mu_b$.

11.10 Collapsed States

If two identical units (a) are connected in parallel, then the Markov diagram of Figure 11.8 can be drawn. Restoration from state 3 has arbitrarily been assigned into state 2. It is assumed that there is only one repairman; otherwise both return arrows from state 3 would be labelled with a μ.

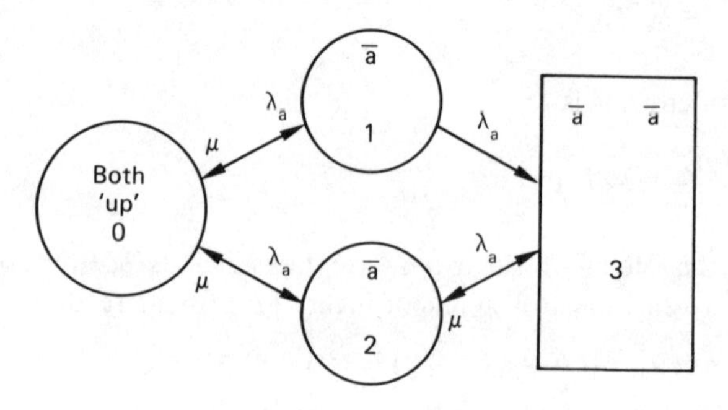

Figure 11.8 Markov diagram for two identical units in parallel

The differential equations derived from this diagram are

$$\frac{dP_0}{dt} = -2\lambda_a P_0(t) + \mu P_1(t) + \mu P_2(t)$$

$$\frac{dP_1}{dt} = -(\lambda_a + \mu) P_1(t) + \lambda_a P_0(t)$$

$$\frac{dP_2}{dt} = -(\lambda_a + \mu) P_2(t) + \lambda_a P_0(t) + \mu P_3(t)$$

$$\frac{dP_3}{dt} = -\mu P_3(t) + \lambda_a P_1(t) + \lambda_a P_2(t)$$

It is possible to collapse states 1 and 2 into a single state $P_c(t)$ since the equations can be re-written as

$$\frac{dP_0}{dt} = -2\lambda_a P_0(t) + \mu[P_1(t) + P_2(t)]$$

$$\frac{dP_1}{dt} + \frac{dP_2}{dt} = -(\lambda_a + \mu)[P_1(t) + P_2(t)] + 2\lambda_a P_0(t) + \mu P_3(t)$$

$$\frac{dP_3}{dt} = -\mu P_3(t) + \lambda_a [P_1(t) + P_2(t)]$$

Putting $P_c(t) = P_1(t) + P_2(t)$ allows the differential equations to be re-written as

$$\left.\begin{aligned} \frac{dP_0}{dt} &= -2\lambda_a P_0(t) + \mu P_c(t) \\ \frac{dP_c}{dt} &= -(\lambda_a + \mu) P_c(t) + 2\lambda_a P_0(t) + \mu P_3(t) \\ \frac{dP_3}{dt} &= -\mu P_3(t) + \lambda_a P_c(t) \end{aligned}\right\} \quad (11.9)$$

These equations are derivable from the Markov diagram of Figure 11.8 in which states 1 and 2 have been collapsed into state c. Figures 11.8 and 11.9 thus represent the same system, but the latter represents a considerable simplification of the former.

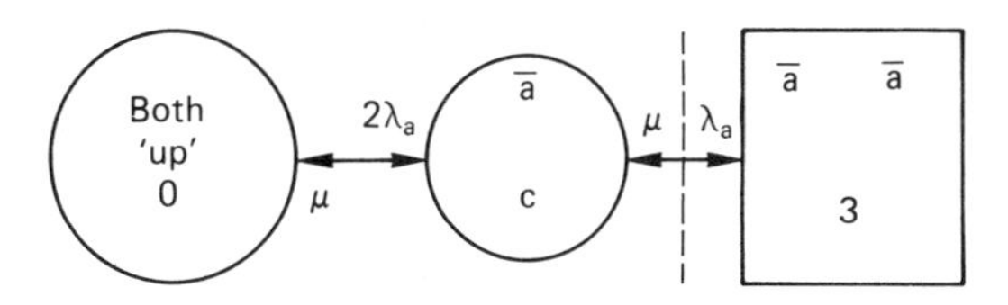

Figure 11.9 Markov diagram with collapsed states

11.11 Parallel Units having an Approximate CFR

The MUT for two identical (repairable) units in parallel can be found from equations (11.9) using the methods of section 11.9.

We need to solve the equations:

$$0 = -2\lambda_a P_0(\infty) + \mu P_c(\infty)$$

$$0 = -(\lambda_a + \mu) P_c(\infty) + 2\lambda_a P_0(\infty) + \mu P_3(\infty)$$

$$0 = -\mu P_3(\infty) + \lambda_a P_c(\infty)$$

with the addition of

$$1 = P_0(\infty) + P_c(\infty) + P_3(\infty)$$

There are four equations in three unknowns, so that one equation can be discarded; the solution is

$$P_c(\infty) = \frac{2\lambda_a}{\mu} P_0(\infty)$$

$$P_3(\infty) = \frac{\lambda_a}{\mu} P_c(\infty) = \frac{2\lambda_a^2}{\mu^2} P_c(\infty)$$

from which the last equation yields

$$P_0(\infty) = \frac{1}{1 + \dfrac{2\lambda_a}{\mu} + \dfrac{2\lambda_a^2}{\mu^2}} = \frac{1}{D} \quad \text{where } D = 1 + \frac{2\lambda_a}{\mu} + \frac{2\lambda_a^2}{\mu^2}$$

$$P_c(\infty) = \left(\frac{2\lambda_a}{\mu}\right) \Big/ D$$

$$P_3(\infty) = \frac{2\lambda_a^2}{\mu^2} \Big/ D$$

so that

$$A_{ss} = P_0(\infty) + P_c(\infty) = \left(1 + \frac{2\lambda_a}{\mu}\right) \Big/ D$$

The rate of crossing the broken line in Figure 11.9 from left to right is $\lambda_a P_c(\infty)$ or $(2\lambda^2/\mu)/D$, so that

$$\text{MCT} = D/(2\lambda_a^2/\mu)$$

Hence

$$\text{MUT} = A_{ss}\text{MCT} = \left(1 + \frac{2\lambda_a}{\mu}\right) \Big/ (2\lambda_a^2/\mu)$$

In many cases, $\mu \gg \lambda_a$ (for example, $\mu = 1/10$ hour^{-1}, $\lambda_a = 1/10\,000$ hour^{-1}), so that we can put MUT $\approx 1/(2\lambda_a^2/\mu)$. It has been shown by simulations that for this combination of units, then so long as $\mu \gg \lambda_a$ the combination has *approximately* a CFR which is equal to 1/MUT or $2\lambda_a^2/\mu$. Corresponding approximations can be made when there are three or more units in parallel.

11.12 More Complex Systems

Markov diagrams for much more complex systems than those shown above may be used for system analysis. The equations involved rapidly become complex algebraically and it is often necessary to use a computer or approximation methods in the analysis. Some tidying-up of the equations may be carried out by using matrix methods.

A particular strength of Markov analysis is that it readily allows the system model to introduce differing repair philosophies as 'repair unit a before unit b' or differing numbers of repairmen, or different restoration rates for different units.

11.13 Constancy of Restoration Rate in Practice

When dealing with real repairmen working in the real world, it might be questioned whether it is reasonable to assume that the restoration rate is constant and the concomitant restoration density function is exponential. In particular, the property of the exponential distribution that the expected time to complete the repair is independent of the time already expended, is unlikely to hold in practice.

Markov analysis using constant restoration rates is very important in practice for three reasons:

(a) it is capable of modelling complex systems using different repair philosophies (such as 'repair x before y');

(b) so long as the restoration rates are high compared with the failure rates (which is usually the case), then simulations have suggested that the exact form of the restoration density function is not critical and that an exponential density function with $\mu = 1/\text{MDT}$ gives results useful in practice;

(c) the method lends itself readily to computer analysis.

Exercises

11.1. Draw a Markov diagram for the system having the Reliability Block Diagram shown in Figure 8.1. Assume that no more failures are possible after a failed state has been reached.

11.2. Calculate the MUT and the MDT using the Markov diagram of Figure 11.7 with the following failure and restoration rates:

$$\lambda_a = 100\,000 \text{ fits}$$
$$\lambda_b = 50\,000 \text{ fits}$$
$$\mu_a = 1/10/\text{hour}$$
$$\mu_b = 1/50/\text{hour}$$

11.3 Calculate the system steady-state unavailability using Figure 11.9. Modify the figure to show the case where there are as many repairmen as faults; calculate the new system steady-state unavailability.

***11.4** A four-engined aircraft can fly on two engines. When all four engines are functioning, the constant failure rate of each engine is λ. If one engine fails, the failure rate of each of the three remaining engines rises to a constant failure rate of 1.2λ. If two engines fail, the remaining two engines have constant failure rates of 1.6λ.

Using Markov analysis calculate the aircraft reliability as a function of λ; you may neglect failures other than those of the engines.

What is the probability (in terms of λ) of completing a seven-hour mission? (Solution of this problem is most easily carried out by the use of Laplace transforms.)

12 Spare Parts Provisioning

12.1 Introduction

In a repairable system it is necessary to hold spare units in stock so that replacements can be made when necessary. Thus, an important part in the planning of system maintenance is the estimation of the number and distribution of repair centres, the number of repairmen stationed at each repair centre and the repair stock which should be held.

In this chapter, methods of calculating the number of spare units which should be held at a repair centre are given. The wider problems concerning the provision of repair centres and repairmen are mentioned below (see Chapter 14). It will be assumed that the systems contain a number of modules* which can be considered to have constant failure rates. A module will generally be a printed circuit board.

Two basic cases will be considered:

(a) failed modules are not repaired so that the repair stock is gradually depleted as repairs are carried out;
(b) failed modules are repaired and eventually returned to the stock room.

In each case, the probabilities of failures of the modules will be assumed to be independent.

12.2 Non-repair of Failed Modules

12.2.1 Identical Modules with CFR

Let the system (or systems) contain a total of m identical modules each of which has a constant failure rate of λ.

Let the probability of not running out of spares during the period of time from 0 to T be P_{UP}. The values of T and P_{UP} will be given: T will normally be the design life of the system and P_{UP} will be some high probability (say 0.99) which reflects how important it is that the system should be kept operational.

*A module may be called a Least Replaceable Unit (LRU).

*12.2.2 Markov Analysis

Analysis can be made using the Markov diagram of Figure 12.1; there is no repair, so the only transitions are from left to right. There are S spares.

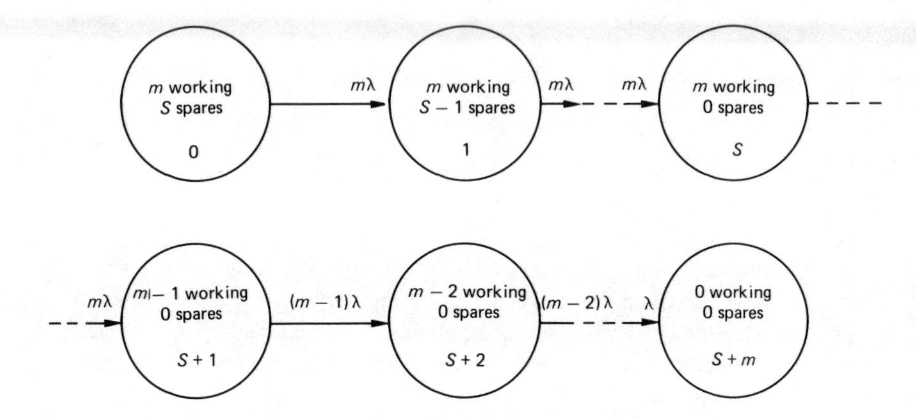

Figure 12.1 Markov diagram for failure of modules without repair

Analysis of the diagram of Figure 12.1 leads to the following differential equations for states 0 to S:

$$\begin{aligned}
\mathrm{d}P_0/\mathrm{d}t &= -m\lambda P_0 \\
\mathrm{d}P_1/\mathrm{d}t &= m\lambda P_0 - m\lambda P_1 \\
\mathrm{d}P_2/\mathrm{d}t &= m\lambda P_1 - m\lambda P_2 \\
&\vdots \\
\mathrm{d}P_S/\mathrm{d}t &= m\lambda P_{S-1} - m\lambda P_S
\end{aligned}$$

Using the nomenclature that $\bar{P}_i(s)$ is the Laplace transform of $P_i(t)$, these equations can be transformed to give

$$s\bar{P}_0(s) - 1 = -m\lambda \bar{P}_0(s) \text{ since } P_0(0) = 1;$$

$$s\bar{P}_1(s) = m\lambda \bar{P}_0(s) - m\lambda \bar{P}_1(s)$$

$$s\bar{P}_2(s) = m\lambda \bar{P}_1(s) - m\lambda \bar{P}_2(s)$$

$$\vdots$$

$$s\bar{P}_S(s) = m\lambda \bar{P}_{S-1}(s) - m\lambda \bar{P}_S(s)$$

which give

$$\bar{P}_0(s) = -1/(s + m\lambda)$$

$$\bar{P}_1(s) = m\lambda\,\bar{P}_0(s)/(s + m\lambda) = m\lambda/(s + m\lambda)^2$$

$$\bar{P}_2(s) = m\lambda\,\bar{P}_1(s)/(s + m\lambda) = (m\lambda)^2/(s + m\lambda)^3$$

$$\vdots \qquad \vdots \qquad \vdots$$

$$\bar{P}_S(s) = m\lambda\,\bar{P}_{S-1}(s)/(s + m\lambda) = (m\lambda)^S/(s + m\lambda)^{S+1}$$

Transformation back into functions of time yields the result (see section 3.16.3):

$$P_0(t) = \exp(-m\lambda t)$$
$$P_1(t) = (m\lambda t)\exp(-m\lambda t)/1!$$
$$\vdots \qquad \vdots$$
$$P_k(t) = (m\lambda t)^k \exp(-m\lambda t)/k!$$
$$\vdots \qquad \vdots$$
$$P_S(t) = (m\lambda t)^S \exp(-m\lambda t)/S!$$

The state probabilities between states 0 and S are thus the terms of a Poisson series with parameter $m\lambda t$.

12.2.3 Required Size of Stock

It has been shown above that the probabilities of failure follow a Poisson distribution, so that in a certain time T the probability that there will be k failures is

$$P_k(T) = (m\lambda T)^k \exp(-m\lambda T)/k!$$

where $0 \leqslant k \leqslant S$ and S is the number of spares. Thus the probability that stock will not be exhausted at time T is

$$P'_{UP} = P_0 + \ldots + P_S$$
$$= \exp(-m\lambda T) + \frac{[\exp(-m\lambda T)](m\lambda T)}{1!} + \ldots + \frac{[\exp(-m\lambda T)](m\lambda T)^S}{S!} \qquad (12.1)$$

In this equation m, λ and T will be given; values of P'_{UP} for given values of S can be found by repeated solution of the equation. Since S must be an integer, it is unlikely that a value of P'_{UP} can be found which exactly equals the required probability P_{UP}. In practice, one thus has two choices:

(1) to take a value of S which gives a value of P'_{UP} closest to the required value P_{UP}; or
(2) to take a value of S which makes P'_{UP} greater than P_{UP}.

As a simple example consider a case with the following parameters:

$m = 100$
$\lambda = 100$ fits
$T = 100\,000$ hours
$P_{UP} = 0.99$

From equation (12.1) it is easily seen that:

$P'_{UP} = 0.920$ when $S = 2$

$P'_{UP} = 0.981$ when $S = 3$

$P'_{UP} = 0.996$ when $S = 4$

Thus to satisfy the required probability it is necessary to have four spares.

It is possible to define the probability of stock-out P_{SO}, that is, the probability that the stock of spares will be exhausted. (However, a request for a spare will only be rejected if another failure occurs after stock-out.) But it should be noted that P'_{UP} and P_{SO} do not add to unity because although

$$1 = P_0 + P_1 + P_2 + \ldots + P_{S+m}$$

$$P'_{UP} = P_0 + P_1 + \ldots + P_S$$

and

$$P_{SO} = P_S + P_{S+1} + \ldots + P_{S+m}$$

so that the term P_S is common to both P'_{UP} and P_{SO}.

12.3 Repair of Failed Modules

When the maintenance arrangements include provision for the repair of failed modules, a more complicated model than that above is needed, since the stock of replacements modules is replenished as failed modules are repaired and returned to the stockroom. The model is then as shown in Figure 12.2.

In the remainder of this chapter, the term 'repair rate' will be used in preference to the 'restoration rate' of Chapter 11, although the same symbol (μ) is used in both cases. This usage serves to differentiate between the conditions considered in Chapter 11 where the repairman travelled to the site of the failure and the conditions assumed here, where the failed module is brought to a repair centre. When the restoration and repair rates can be considered to be constant (or approximately so) their inverses are the MDT and the Turnaround Time respectively. In US literature, the Turnaround Time may be referred to as the Spares Replacement Interval (see Jeschke *et al.*, 1982, sections 5.7 (Sparing Philosophy) and 5.8 (Sparing Strategies)).

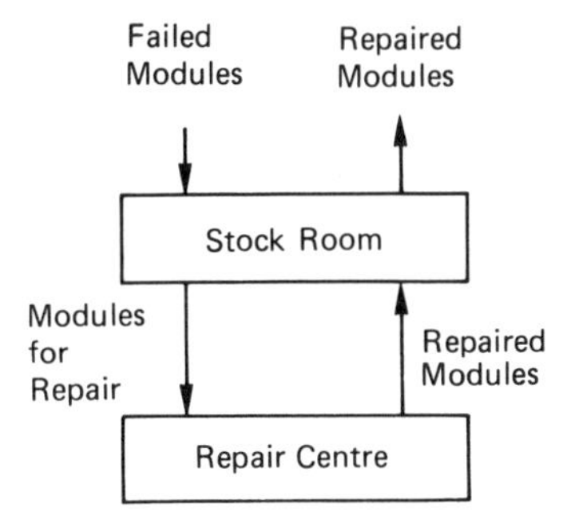

Figure 12.2 Provision of a repair centre

It is convenient to consider that the repair rate (μ) of failed modules is constant. Although this may not be true in practice, the inaccuracies involved in this assumption are usually acceptable (see section 11.13). It should be noted that this repair rate refers to one repairman working upon one failed module. It is assumed that even if more repairmen are available, they will not combine with the first man so that the repair rate per module remains the same. However, if r repairmen are available, then the repair rate for $1, 2, 3 \ldots r$ failed modules is μ, $2\mu \ldots r\mu$. The repair rate must remain at $r\mu$ however many more modules fail, since the $(r + 1)$th, $(r + 2)$th . . . failed module must be neglected until a repairman becomes available.

The Markov diagram is shown in Figure 12.3, relating to the position when there are m modules used in the system, each of which has constant failure rate λ; there are S spare modules; there are r repairmen; the repair rate for a module is μ; $S > r$. The figure should be re-drawn and re-analysed if these conditions do not apply.

State $(r + 1)$ is the state at which there are not enough repairmen to cope with the number of failures. At state S, stock-out has been reached, although all m modules are working. From state $(S + 1)$ onwards, there is a steadily decreasing number of working modules.

If P_i is used to denote the steady-state probability that the system is in state i, then steady-state equations can be written down as follows:

$$\begin{aligned} m\lambda P_0 &= \mu P_1 \\ m\lambda P_1 &= 2\mu P_2 \\ &\vdots \\ 2\lambda P_{S+m-2} &= r\mu P_{S+m-1} \\ \lambda P_{S+m-1} &= r\mu P_{S+m} \end{aligned}$$

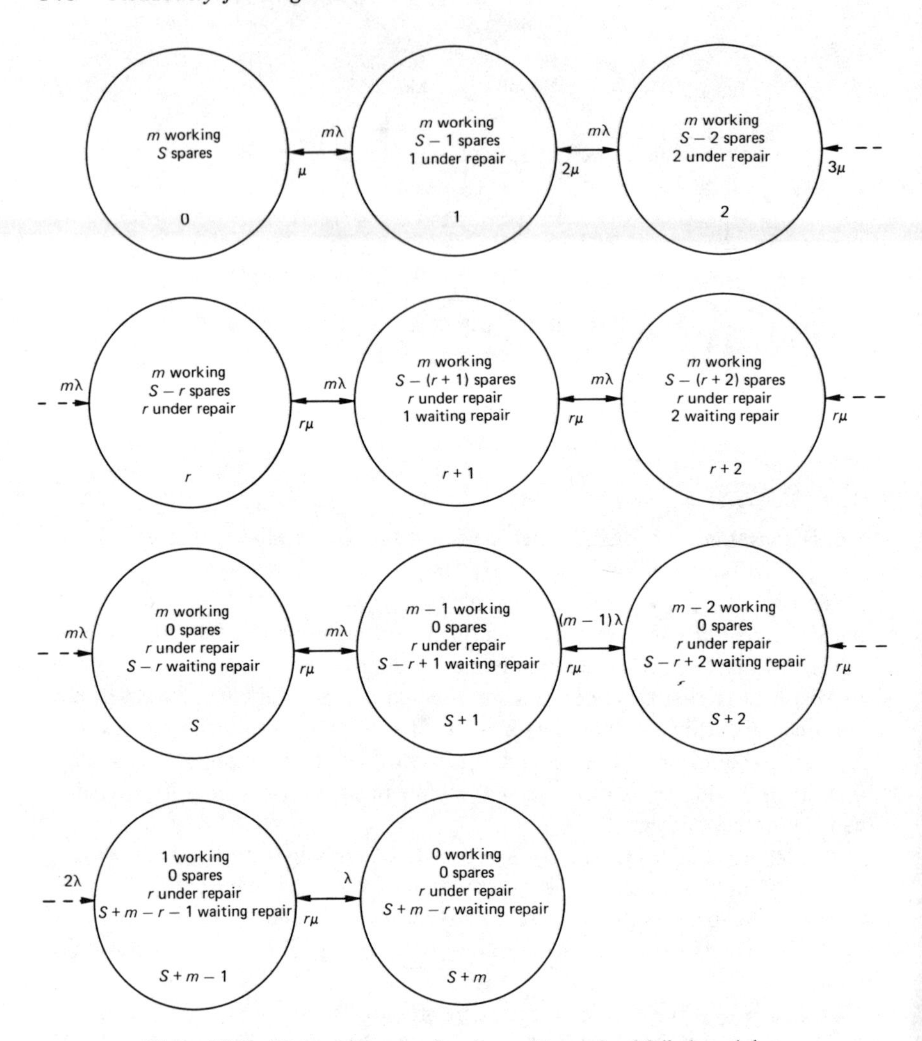

Figure 12.3 Markov diagram of system with repair of failed modules

By finding $P_1, P_2 \ldots P_{S+m}$ in terms of P_0 and using the relation

$$1 = P_0 + P_1 + \ldots + P_{S+m}$$

the set of steady-state probabilities may be obtained. The procedure is straightforward, but the equations are somewhat complex. A simplification which is often used is to assume that there are as many repairmen as faults.

The ith state in Figure 12.3 can then be labelled:

$$m \text{ working}, \quad (S - i) \text{ spares}, \quad i \text{ under repair for } 0 \leqslant i \leqslant S$$

and the $(S+j)$th state can be labelled:

$(m-j)$ working, 0 spares, $(S+j)$ under repair for $1 \leqslant j \leqslant m$

The return rates between states are

μ from state 1 to state 0
2μ from state 2 to state 1

$\vdots$

$(S+m)\mu$ from state $(S+m)$ to state $(S+m-1)$

Solving these equations by methods described above yields:

$$\text{putting } D = \sum_{i=0}^{S} \left(\frac{m\lambda}{\mu}\right)^i \frac{1}{i!} + \sum_{j=1}^{m} \left(\frac{m\lambda}{\mu}\right)^S \frac{m(m-1)\ldots(m-j+1)}{(S+j)!} \left(\frac{\lambda}{\mu}\right)^j$$

then

$$P_i = \left(\frac{m\lambda}{\mu}\right)^i \frac{1}{i!} \Big/ D \quad \text{for } i = 0 \text{ to } S$$

$$P_{S+j} = \left(\frac{m\lambda}{\mu}\right)^S \frac{m(m-1)\ldots(m-j+1)}{(S+j)!} \left(\frac{\lambda}{\mu}\right)^j \quad \text{for } j = 1 \text{ to } m$$

The state probabilities may be calculated manually but the computation is most easily done using a computer and custom-written or commercially-available software (see section 19.3).

A number of useful parameters may be introduced following the analysis of Figure 12.3, chief of which is probability of stock-out which is

$$P_{SO} = P_S + P_{S+1} + \ldots + P_{S+m}$$

where P_i is the steady-state probability that the system is in state i.

It should be noted that the probability that all m modules are up and running is not $1 - P_{SO}$ (see also section 12.2) since

$$P'_{UP} = P_0 + P_1 + \ldots + P_S$$

and

$$P_{SO} = P_S + P_{S+1} + \ldots + P_{S+m}$$

so that P_S occurs in both expressions.

In general it will be possible to make trade-offs between the various provisioning parameters. Thus, if m and P_{SO} are given, then it is possible to obtain a table of solutions for differing values of μ (repair rate), r (number of repairmen) and S (number of spares).

It has been tacitly assumed in the above treatment that the failure rate of a module during storage is zero. This is not necessarily the case and provision for this possibility can be made. See Pullum and Grayson (1979).

The analysis of spare parts provisioning is usually concentrated upon the steady-state probabilities of the system states. However, if circumstances demand otherwise, the dynamic behaviour of the probabilities (that is, as functions of time) can be analysed using the principles of Markov analysis described in the previous chapter.

Exercises

12.1. Calculate P_{SO}, the probability of stock-out, for a non-repairable module using the following parameters (see section 12.2):

$$m = 10, \quad \lambda = 1/1000/\text{hour}, \quad T = 200 \text{ hours}, \quad S = 2$$

12.2. Draw a Markov diagram, similar to Figure 12.3 for the case when

$$m = 3, \quad S = 4, \quad r = 2$$

13 Software Reliability

13.1 Introduction

Although the term 'software reliability' is widely used, it is something of a misnomer since any software errors (or bugs) which are in a program are there at the time the system is released and remain in the program until they are removed by re-programming. For this reason 'software design adequacy' (as suggested by Dr G. G. Pullum) is really preferable to 'software reliability'.

The reader may question why it is that software should ever fail. Is it not possible to produce a perfectly logical piece of software which has no errors whatsoever? In theory, the answer must be 'yes'; in practice, any program of reasonable size has so many possible paths through it that it is impossible to test or check every one. As a result, systems containing software can (and do!) fail fairly regularly.

Work is being carried out on ways of proceeding directly from a formal logical specification to a software program; the problem here is that it is generally very difficult to write a completely water-tight system specification (that is, one that defines exactly what the system should do under all possible circumstances). It is not unusual, during system development, for a designer to ask the question 'what should happen if . . .?' and to find that the answer is not in the specification, because the problem had not been anticipated when the specification was drawn up. As a result of inherent errors in the software, systems fail (that is, cease to operate properly) at times which are effectively randomly distributed. As with hardware failures, methods are needed for the prediction of times to failure.

In some ways it would have been better to use the term 'software bug' rather than 'software error', since 'error' can be used to mean a number of different things. However the term 'bug' has not been used since 'bug avoidance' and 'bug tolerance' may have produced the wrong sort of pictures in readers' minds!

The differences between hardware failures and software errors are discussed in section 13.2. The following section ('The Production of Reliable Software') briefly mentions the most obvious ways of avoiding software errors and mitigating their effects. 'Some Aspects of System Design' (section 13.4) examines the interaction that reliability problems produce in hardware and software design. Finally, section 13.5 discusses how software reliability can be predicted.

For more information on the vast subject of software reliability, the reader is referred to specialised texts like Myers (1976) or Musa *et al.* (1990).

13.2 Differences between Hardware Failures and Software Errors

In conventional reliability theory, the failure of components is considered to be caused by either excessive parameter drift or catastrophic breakdown. Whatever the cause of them may be, failures are assumed to occur randomly in time, and the failure distribution of the vast majority of electronic components is considered to be exponential. In electronic systems the repair of hardware, following failure, will put the system back in the 'as-good-as-new' state so long as the components used follow an exponential failure law (or equivalently, have a constant failure rate). This useful attribute simplifies the analysis of hardware systems and enables the powerful Markov analysis procedures to be used.

System failures caused by software errors are of a completely different nature from those caused by hardware failures. They cannot be said to occur randomly in time since the errors must be considered to be present in the program from the very beginning. It can, however, be argued that they show up at random points in time. However, the philosophical basis for the claim that the appearance of software errors follows an exponential law seems to be questionable. Moreover, when a software error is 'repaired', the system enters a previously unknown state: the system is not 'as-good-as-new' as it was following a hardware repair, and Markov analysis of such behaviour can rapidly become unwieldy.

The types of transition which can occur as a result of hardware failures or software errors are shown diagrammatically in Figure 13.1. A hardware failure will cause the state to change from the initial state to one of a number of possible states (n, o, p, . . .) which can be anticipated and which can be enumerated by means of a Failure Mode, Effect and Criticality Analysis (FMECA). The effect of the interaction between the software and the failed hardware may produce yet other system states (n′, o′, p′, . . .) which may or may not be anticipated.

These states have thus been labelled 'Possibly Anticipated States'. Similarly, the entry of invalid data will cause the system to enter states j′, k′, . . . which may not be anticipated. The effect of software errors will be to cause the system to enter unanticipated states (a′, b′, . . .); defensive programming can be used to alleviate some of the problems associated with entry of the system into these states. Transient failures will also cause the system to enter some unanticipated states (f′, g′, . . .). System restart may then often be the most convenient method of returning the system to its initial (working) state.

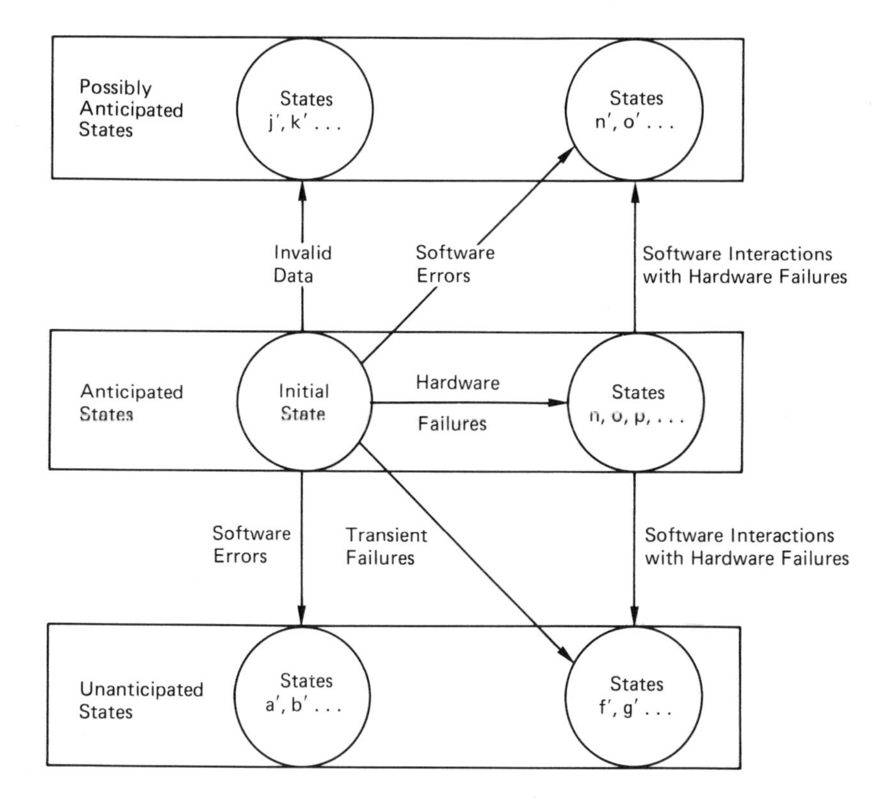

Figure 13.1 Possible system states following the appearance of hardware failures or software errors

13.3 The Production of Reliable Software

13.3.1 General

There are three areas in which reliability considerations affect the production of software:

(a) Error avoidance. The software should be as error-free as possible.
(b) Error tolerance. Taking into account the necessary financial and time constraints in a project, consideration should be given to making the software error-tolerant, so that when a software error is encountered the system does not fail completely.
(c) Graceful degradation. Those errors which do occur and cannot be tolerated should be made to have as little effect on the system as possible (for example, by giving a degraded service rather than none at all).

Item (a) above can be considered in isolation from the hardware. However, the implementation of items (b) and (c) may have an impact upon the hardware design (for example, redundant hardware units may be needed to implement error tolerance and graceful degradation).

'Error' avoidance and 'error' tolerance are used here instead of the more usual 'fault' avoidance and 'fault' tolerance in order to differentiate between hardware failures (which occur at specific times and produce faults) and software errors (or bugs) which are present from the moment that the software is released to the customer.

13.3.2 Error Avoidance

Methods of trying to avoid the occurrence of errors in software before it is released may be broadly based under three headings:

(a) The organisation of the software itself and of the team producing it.
(b) The writing of the software.
(c) The testing of the program before it is released.

In (a), good methodologies for structuring the software and the team of programmers is essential, and these are widely reported in the literature.

To achieve a minimum number of errors in the actual writing of the program [(b) above], various formal methodologies have been suggested (such as the Vienna Development Method). Semi-automatic writing methods have been proposed but they do not yet appear to have been very successful.

Many methods of program testing are available and literature on this topic is widely available.

A system failure which is caused by unacceptable input data (such as a negative signal power) should not be regarded as being caused by a software error (as defined here). Nevertheless, it is obviously desirable for the software to contain traps for this type of problem and for the system to behave in a defined manner when it occurs. The term 'defensive programming' is used to describe such procedures.

13.3.3 Error Tolerance

'Error tolerance' can have a variety of meanings. Here 'error' is taken to mean a mistake in the software (a software bug) which could, at least in theory, be eradicated.

Various methods of error tolerance in software alone have been suggested. An example is the technique using 'recovery blocks' (Randell (1975)); when a malfunction is detected in the system, the program goes back to a previous point and tries an alternative method of computation. The extra time needed for recomputation is a disadvantage in this method, which may not be suitable for some real-time systems.

Mixtures of hardware and software can be fault tolerant as, for instance, in the system shown in Figure 13.2 in which two units are in parallel redundancy. The two parallel units use different hardware (H_a, H_b) and software (S_a, S_b); so long as the errors in the one set of software are not contained in exactly the same way in the other, then the system will have a certain amount of error tolerance.

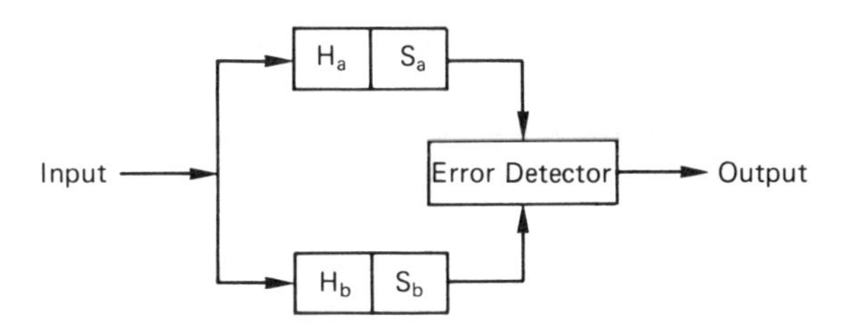

Figure 13.2 A system with redundant hardware and software

13.3.4 Graceful Degradation

This is sometimes known as a fail-soft mode of failure. If a system is designed so that most hardware failures and software errors cause partial but not total loss of system functions, then it is said to degrade gracefully. Such graceful degradation is often attractive to users since in many cases reduced or restricted performance is preferable to no performance at all. (Most telephone exchanges are designed to degrade gracefully.) The topics of graceful degradation, redundancy, defensive programming and software module size are all inter-linked.

13.3.5 Prediction of Software Reliability before Start of Development

Most models for predicting software reliability rely upon some method of extrapolating from the rate of their appearance during system development. It may happen that when tendering for a contract, before development has actually started, that no such information is available. This is a difficult problem to solve; the reliability engineer may have to use past data (if available) plus his engineering judgement. The problem is discussed in Kitchenham (1987).

13.4 Some Aspects of System Design

13.4.1 System Design and Development

In the design stage of a system, it is usually necessary to carry out a number of iterations before an acceptable design finally emerges. For each iteration, it is necessary to predict both the reliability and the development costs involved along with many other factors. There is a vicious circle in that the prediction of programming effort required for the project depends (at least to some extent) upon the reliability predictions (for example on how much redundancy is required) while the reliability predictions will depend (at least to some extent) upon the programming effort available.

The only way to break this vicious circle is for the design team to make a guess of (say) the programming effort required, and then to examine whether a consistent design and development plan emerges from the calculations. Such a procedure calls for a good deal of finesse on the part of the design team, and clearly software reliability adds significantly to the problems involved.

13.4.2 Hardware 'versus' Software in System Design

Over the past 15 to 20 years, the cost of software in a 'typical' system development has risen from about 15 per cent to about 80 per cent. There are two principal factors which have brought this change about:

(a) Decreasing cost of hardware brought about by higher circuit complexity.
(b) Increasing cost of software caused by increasing complexity and difficulty in 'debugging'.

During the period of time considered, the trend has been for software to be used to replace hardware wherever possible. There are good reasons for this trend: software is inherently more flexible than hardware, and can be updated and improved simply (although the simplicity may be illusory in a complex system!); using a minimum of hardware reduces the physical size, physical complexity and power consumption of the system.

It may be questioned whether the trend mentioned above should be reversed. Squeezing a quart into a pint pot has led to very high software costs and to systems which seem to be almost inherently unsatisfactory. The advantages to be gained by using hardware in at least some of the places where software could be used are:

(a) Hardware is usually simpler to design and test than software. (It must be conceded, however, that it is proving difficult to test some VLSI chips adequately.)

(b) Hardware interfaces are more easily defined than software interfaces, and are thus less likely to pass corrupted or incorrect information.
(c) Reducing the size and complexity of software modules should result in fewer software errors, easier testing and verification, and consequent cost reduction.
(d) Owing to the high cost of testing and maintaining software (up to 75 per cent of total software costs - see Myers (1976)) it is likely that a system's life cycle cost could be cheaper if hardware were used in at least some of the places where it is optional. (It must be said, however, that 'maintenance' here does include costs of changing the software to comply with altered system requirements.)

13.4.3 Distributed Control

The advantages of distributed processing should always be considered. This does not necessarily mean replacing software by hardware, but rather using several small processors instead of a single large one and placing well-defined interfaces (preferably hardware) between them. Distributed processing allows graceful degradation of the system and reduction in size of software modules (leading to cheaper and more reliable software).

13.4.4 Size of Software Modules

It can be argued that reliable large software programs can be constructed by interconnecting many small (and hence reliable) programs in a well-defined structure; this is the basis of many software architecture designs. However, in practice it still seems to be the case that software remains unreliable. It thus appears that in many applications it is best to use small well-defined software modules interconnected through hardware interfaces. Distributed processing, graceful degradation and high software reliability should result from the use of such architectures.

13.4.5 Error Detection and Correction

This term can cause some confusion when it is applied to software. In normal telecommunications usage the term is applied to a method (which can be embodied by hardware and software) to detect errors in a data stream. Redundant codes are used; the simplest form of error detection is the use of an extra parity bit per byte of data. The errors which occur in a data stream may indicate a hardware failure, a software error or a transient condition caused by electrical noise.

It seems best to avoid the use of the term 'error detection and correction' when applied to software. Software can detect only an unacceptable set of parameters: the 'error' may have arisen through unacceptable input data, a hardware failure or a 'bug' in the software itself. It seems unlikely that software can ever 'correct' itself since this would involve the software actually rewriting itself. 'Error avoidance' can be carried out by software since if a faulty software module is identified, the program may contain instructions to avoid the use of that particular module.

13.5 Software Reliability Prediction

13.5.1 Appearance of Software Errors

A problem in the prediction of software reliability is caused by the regular updating of software. When software is first issued (as 'Issue 1'), the rate of appearance of errors is usually high initially and then starts to fall off, as shown in Figure 13.3.

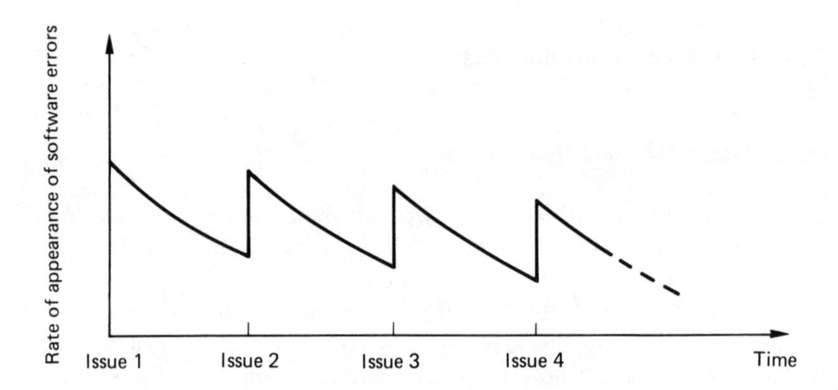

Figure 13.3 The appearance of software errors as a function of time

Errors which have been found in the software are corrected, often in a temporary fashion. After a certain time, the software is re-written and released as Issue 2. In the re-writing, although previously-discovered errors are removed, often new errors are introduced. The effect of this is to bring about a sharp increase in the rate of appearance of software errors. The rate then decreases until a subsequent Issue causes a further increase, as shown in Figure 13.3. This process can continue indefinitely, although it is to be hoped that the number of errors (and their rate of appearance) does tend to diminish. However, the saw-tooth appearance of Figure 13.3 does indicate a severe difficulty in software reliability prediction.

13.5.2 Problems

It has been shown above (section 13.2) that there are basic differences between how hardware failures and software errors manifest themselves. The modelling of system failures caused by software errors has proved to be particularly difficult. A survey made in 1987 found 41 different models - the number is still probably increasing. No particular model has yet achieved anything like universal acceptance.

Some of the problems of modelling software reliability are:

(a) The incidence rate changes with updates of the software, as shown in Figure 13.3.
(b) Changes in the programming team during the development period can cause the rate of appearance of errors to change (for better or worse!) and thus a model which depends upon extrapolation of data may become inaccurate.
(c) Programmers may be unwilling to provide data upon the number of errors they introduce into programs since they fear that the data may reflect upon their competence.

13.5.3 Solutions

Despite the problems stated above, the reliability engineer must make some attempt at providing a software reliability prediction even if it is only a 'guesstimate'. Even though there are fundamental differences between hardware failures and software errors (see section 13.2) it is usual to assess software reliability in terms of the functions which have been already developed in this book, that is, as an equivalent reliability ($R_{SW}(t)$) or failure rate* ($\lambda_{SW}(t)$). The resulting value can then be incorporated with that for the hardware reliability as explained in Chapter 15.

Suggested methods are:

(a) Assess the software failure rate as a proportion of the hardware failure rate. Wherever possible, past data on similar system developments should be used for estimation purposes. In the total absence of previous data, a value of 30 per cent is tentatively suggested, although engineering judgement must always be used, since in a system or subsystem containing much software, the quoted value may be too low. Some data on this topic are available in the Bellcore (Bell Communications Research, USA) Document 'Proposed Revision of TR-TSY-000512, Reliability Performance Objectives for LATA Switching Systems' reference TA-TSY-000512.

*Musa *et al.* (1990) use the term 'failure intensity' rather than 'failure rate'. In this context the two terms can be considered to be identical. (In general, 'failure intensity' is applied to repairable systems and it has not been found necessary to use it in this book.)

(b) Use an existing software reliability model to assess the software reliability ($R_{SW}(t)$) or failure rate ($\lambda_{SW}(t)$). Suggested models are those of: (1) Shooman (see Roberts (1988)); (2) Littlewood–Verrall (see Littlewood and Verrall (1973)); and (3) Musa–Okumoto (see Musa and Okumoto (1984)).

Examples of the use of software reliability functions for system reliability prediction are given in section 15.6.

13.5.4 Some Recommendations on Software Reliability

Broadly-based recommendations for the planning and writing of software are:

(a) Define the architecture (that is, inter-relationships between different software modules) carefully and make it as simple as possible.
(b) Use a 'top-down' approach in program planning, starting at the least detailed level and then working downwards in increasing detail.
(c) Carefully control all documentation and ensure that adequate, clear descriptions of the architecture and the program modules are provided.
(d) Separate the program into different functional modules.
(e) Where possible, use defensive programming; make provision (if possible) for automatic re-start after a software 'crash'.
(f) Use established Quality Assurance procedures, for example those published by the Ministry of Defence (UK) in Defence Standard DEF–STAN 00–16. See also Cho (1987).
(g) Write the program using simple rather than complex programming techniques.

The last recommendation (g) may seem strange. But it has been found in the past that if a clever programmer writes a difficult-to-understand (albeit short) set of instructions, then a later programmer may have so much difficulty in trying to understand and adapt it (given that software maintenance is required) that the whole set of instructions must be re-written. Since there are always changes taking place in programming teams, it is important that each programmer should (1) write simple, clearly-understood programs and (2) provide an ample description of what he has done.

14 Maintainability

14.1 Introduction

Maintainability is defined as 'the probability that a repair action will restore a failed system to full working order within a given period of time'. It is thus closely analogous to reliability since both are probabilities which vary with time. The chief difference conceptually between maintainability and reliability is that in general there are more trade-offs that can be made in the former (number of repairmen, skill of repairmen, number of repair centres, number of modules held in repair stock, provision of alarms in the system, maintenance schedules etc.) than in the latter.

It is an important feature of system development that ease of repair and provision of fault alarms should be an integral part of the initial design. An approach which attempts to 'bolt on' maintenance to a nearly fully-designed system is to be deprecated.

14.2 General Maintenance Problems

Some of the general problems in providing maintenance are:

(1) The provision of Repair Centres:
  - (a) how many should there be?
  - (b) where should they be situated geographically?
  - (c) how many repairmen should be allocated to each centre?
  - (d) how many spares should be held at each centre?

(2) The provision and effectiveness of alarms:
  - (a) what failures should be alarmed?
  - (b) how should urgent and non-urgent alarms be allocated?
  - (c) how should on-site and remote alarms be allocated?
  - (d) how should the man–machine interface be designed?
  - (e) which failures* will not cause an alarm to be raised?

*The term 'fault coverage' is sometimes used: it is defined as the percentage of total possible faults that a system can detect.

(f) to what level (for example, PCB or subsystem) should alarms identify faults?

(3) The provision of built-in test equipment (BITE).

(4) The allocation of maintenance times (see section 10.2.1):
(a) response delay time
(b) travelling time
(c) active repair time
(d) post-repair administrative time.

(5) Skill of the repairmen.

14.3 Preventive Maintenance

Preventive maintenance is important in three main areas:

(a) Replacement of components or modules whose wear-out occurs before the service life of the system is reached (for example, electrolytic capacitors).
(b) Routine maintenance of components (for example, oiling or greasing of bearings) in order to reduce the probability of failure before the expected service life.
(c) Routine checks on dormant failures (see section 16.4.5).

Generally, electronic components have useful lives which can be considered to be well in excess of system service lives. However, there are exceptions (electrolytic capacitors, microwave tubes etc.) which do have limited lives. Electro-mechanical devices (such as cooling fans) may need to be replaced some time before wear-out is expected, in order to ensure that the failure rate remains acceptably low. So long as these components are routinely replaced as part of the maintenance plan, then it is acceptable to treat them as constant failure rate devices (see section 10.3).

Preventive maintenance is usually expensive to provide and there is a strong tendency to phase-out devices like adjustable electro-mechanical relays (as used in the old 'Strowger' telephone exchanges).

14.4 Corrective Maintenance

It was explained in section 10.3 that the distribution of downtimes may not be exponential in practice; some authorities claim that downtimes are likely to be log-normally distributed. The distribution of downtimes can be an important feature in system maintenance although it is most often sufficient to consider the mean downtime; this will generally consist of the sum of the means of travelling time, administrative time and active repair time (see section 10.2).

14.5 Maintenance and Spare Parts Provisioning

The probability of stock-out (see Chapter 12) should be assessed, taking the following factors into account:

(a) The cost of one or more systems going down and staying down while more spares are procured.
(b) The possible long delay if more spare parts have to be procured. (Specialised spare parts may be unobtainable after an initial production run.)
(c) The importance of keeping good customer relations.

14.6 Trade-off between Reliability and Maintenance

It is generally possible to give a basically unreliable system a high availability by decreasing the downtime. Owing to the high cost of maintenance engineers (particularly at weekends!) there is a strong movement towards making systems more reliable and hence reducing the cost of maintenance. Methods of increasing the system reliability and maintainability include the use of redundancy, self-diagnostics and remote alarming.

14.7 Software Maintenance

Good Quality Assurance procedures on documentation and program structure should be followed during system development. If this is done, then when software maintenance is required, it should be possible for the same programmer (or more probably a different programmer) to understand and modify the program properly.

14.8 Provision of Maintenance Manual

A system Maintenance Manual should describe:

(1) Preventive maintenance actions.

(2) Corrective maintenance actions, especially:
  (a) the provision of urgent and non-urgent alarms;
  (b) the sequence of steps to be taken in finding a fault;
  (c) special instructions (for example, it may be necessary to remove power from a PCB before extracting it).

(3) Skills which are required of a repairman.

(4) Equipment required for carrying out each repair.

14.9 Demonstration of Maintainability to a Customer

Some customers (such as PTTs) ask a manufacturer to demonstrate the maintainability of his system. The demonstration consists of the customer putting a number of known faults in the system and then noting the time taken to locate and repair the fault. The procedures used for fault location and repair should follow those given in the Maintenance Manual, and the repairmen should have the abilities prescribed in the Manual.

14.10 Allocation of Maintainability

During system design, it may be necessary to allocate different maintenance times to different subsystems so that the total system reliability parameters are achieved. The allocation of maintainability follows the principles given in the allocation of Reliability in Chapter 16.

14.11 Cost of Maintenance

14.11.1 Overall Costs

During system planning it may be necessary to estimate the cost of system maintenance. For electronic systems which are well-maintained and used in a benign environment, the cost of maintenance per year should be in the region of 5-10 per cent of the capital cost. For systems used in non-benign environments (for example, military systems), the cost of maintenance per year may be as high as 15-20 per cent of the capital cost. These are total costs which allow for the costs of spares and labour costs of repairmen. They do not include any capital costs incurred (as, for example, the building of repair centres).

14.11.2 Number of Repairmen Required

In calculating the number of repairmen required to carry out a certain maintenance activity, the following provisions for absence should be made:

(a) holiday;
(b) sickness;
(c) training (as an on-going activity);
(d) miscellaneous absences (jury service, for example).

Although overtime can sometimes be used to make up for deficiencies in manpower, it is probably better not to include this as anything but a temporary part of the manpower plan.

As an example of manpower planning, it will probably be necessary to employ five or six repairmen to provide one repairman who is on call for 24 hours a day on every day of the year.

14.12 Integrated Logistics Support

This consists of the comprehensive planning of maintenance through the design, production and field use of (usually military) equipment. It requires the co-operation of designers, maintenance planners and the customer's representative to ensure that from the initial stages of planning through to use in the field, all aspects of design and maintenance and their influence upon the life cycle cost are taken into account. Aspects which are considered include:

(a) corrective maintenance;
(b) preventive maintenance;
(c) spare parts ranging (providing a list of spares);
(d) spare parts scaling (provisioning);
(e) documentation (maintenance manual, flow diagrams etc.);
(f) training of maintenance staff;
(g) selection and provision of tools and test equipment;
(h) building and organising of repair centres;
(i) life cycle cost.

14.13 Maintenance Checklist

A checklist is given below which can be used in trying to ensure that no aspects of maintenance are neglected. The same item may appear under several different headings; in this way it is hoped that the inter-relationship of various factors in maintenance will be highlighted.

Five phases in the life of a system have been taken as typical, and headings are grouped under these phases as appropriate.

DESIGN AND DEVELOPMENT PHASE

1. *Equipment*
   1.1 Ease of maintenance of the hardware:
      type of equipment practice
      type of interconnection
      replacement of back-plane
      replacement of fans
      replacement of fixed equipment (gas detectors, heaters etc.)
      self-diagnostics
      provision of test equipment (built-in and portable)
      use of standard or custom-built test equipment

1.2 Ease of maintenance of the software:
software architecture
software documentation
provision of RAM, EPROM, EEPROM etc.
ease of recovery and restart
programming language used

1.3 Dormant failures:
identification of FMECA by field failure data (event reporting)
effect on preventive maintenance

1.4 Design rules:
overstressing of components
complexity and design of printed circuit boards
man–machine interfaces
equipment practice
second-sourcing of equipment and components

1.5 Provision and effectiveness of alarms:
what failures should be alarmed
allocation of urgent and non-urgent alarms
design of man–machine interface (see also below)
'fault coverage' (percentage of total possible faults which are detected)
level to which alarms identify faults (pcb, rack etc.)

1.6 Test equipment:
built-in
carried by repair man (portable)
special to the system (custom-built)
non-special to the system (standard)

1.7 Maintainability:
allocation of downtimes to various subsystems or modules

1.8 Service life:
desired service life of system

1.9 Safety:
flammability of components and materials
hazards to operating and maintenance staff

2. *Documentation*
maintenance manual
failure dockets (to be filled in by repairman and used in event reporting)
operating manual
instruction manual (if different from operating manual)
installation manual

3. *Economics*
life cycle costing
ratio of (capital cost)/(cost of ownership)
yearly maintenance cost
one-off continuous purchase of spares

cost of repairmen, repair centres, spares holdings of equipment and components
cost of alternative maintenance strategies

4. *Man–machine interface*
   4.1 Alarms
       number and type of alarms
       allocation of urgent/non-urgent alarms
       alarm display (message on VDU, lamp, bell etc.)
   4.2 Human factors
       ergonomics of maintenance (such as ease of removal of equipment from cabinets)
       ease of use of portable and built-in test equipment
       equipment practice
       alarm strategy

MAINTENANCE PLANNING PHASE

5. *Repair strategy*
   5.1 Uptimes:
       calculation of system, subsystem and module reliabilities
   5.2 Downtimes:
       assumed time to identify fault
       assumed travelling time
       provision and siting of repair centres
       spares holding (spare parts provisioning)
       skill of repairmen
       equipment, spares etc. to be carried by repairman
       training of repairmen
       maintenance manual
   5.3 Provision of Repair Centres:
       number and spares holding at each (pcbs or LRUs)
       siting
       travelling times
       number of repairmen at each site
   5.4 Equipment repair strategy:
       repair/scrap decisions
       spares holdings of replacement components
       second-sourcing of equipment and components
       provision for repairs (on site, return to manufacturer etc.)
6. *Training*
   training of repairmen (including failure reporting or event reporting)
   assumed skill levels of maintenance staff
7. *Spare parts provisioning*
   failure rates, turnaround time, problem of stock-out

second sourcing
once-off or continuous purchase of spares

PRODUCTION PHASE

8. *Quality*
component quality
manufacturing quality (for example, quality of soldering process)

INSTALLATION PHASE

9. *Demonstration test*
demonstration of maintainability by means of a test
10. *Installation*
skill of installation staff
installation manual
training of installation staff

OPERATIONAL PHASE

11. *Operation*
skill of operating staff
operating manual
training of operating staff
12. *Event reporting (follow up)*
analysis of reported failures
analysis of repair actions, mean downtimes etc.
identification of critical components
feedback of information to design team
evaluation of achieved system and subsystem reliability and maintainability
identification of necessary design, procurement or manufacturing changes
identification of improvements needed in maintenance (maintenance procedures; diagnostic aids; documentation; test equipments; training etc.)
13. *Preventive maintenance*
frequency of routine inspection (such as for physical damage, like cracks in cabinet)
inspection for dormant failures
routine replacement of wear-out items (fans etc.)
14. *Corrective maintenance*
frequency of corrective maintenance actions
cost of corrective maintenance
maintenance manual
repair/scrap strategies

15 System Reliability Prediction

15.1 Introduction

In this chapter the methods of reliability prediction which have been described in previous chapters are summarised and a consolidated process for system reliability prediction is presented.

15.2 The Process of System Reliability Prediction

The process usually consists of the following steps:

(a) Identification of the important system failure modes.
(b) System reliability analysis for each important failure mode. Using the methods of Chapters 6 to 11, the required parameters (MTTF, MUT, Availability, Mission Reliability etc.) are calculated in the form of algebraic functions of hardware and software failure rates and restoration rates (if these are required).
(c) The choice of suitable hardware failure rate data (for example, HRD4) together with numerical values of the software failure rates and restoration rates where needed.
(d) Calculation of the numerical values of the reliability parameters using the algebraic functions from step (b) and the numerical data from step (c). Reference should be made at this stage to the Mission Profile to ensure that the failure rates being used are compatible with the environments which the system will encounter.

Each of these steps will now be examined in more detail.

15.3 Identification of System Failure Modes*

This topic was introduced in section 6.2. In general there will be both hardware and software failure modes; the most important of these can often be identified by asking the question 'What does the system do?', and following this by asking 'In what ways can it cease to perform these functions?' Thus the cassette radio described in sections 6.2 and 6.4 has at least the following failure modes:

failure of the radio only
failure of the cassette player or recorder only
complete failure.

Other failure modes could be identified, for example 'failure of one stereo channel'.

More complex systems than the cassette radio are likely to contain significant amounts of software and these will produce their own failure modes. Software failure modes may be more difficult to predict than hardware failure modes (if they can be identified, then the software can possibly be modified so as to remove them completely; those that remain are difficult to identify). To defend against software errors, some of the fault-tolerant methods described in Chapter 13 can be used. Thus it may be possible to design the system to have automatic re-start after a system software crash. In these circumstances, whether or not a system software crash is defined as a system failure must depend upon the use to which the system is put.

In some cases, possible failure modes are not easy to define and recourse must be made to Failure Mode, Effect and Criticality Analysis (FMECA), see Chapter 19. Such an analysis is particularly desirable when it is known or suspected that a system can have hazardous failure modes.

15.4 System Reliability Analysis

For simple series systems, the reliability analysis may consist of no more than adding component failure rates and (where necessary) software failure rates. However, care must always be taken in this apparently simple procedure, to ensure that only those components whose failure rates correspond to the chosen failure mode are included. For example, a series system may contain alarm circuitry, failure of which does not cause system failure; in this case the failure

*It would be difficult to over-emphasise the importance of this step as is evidenced by the following case history. A failure, whose mode was unpredicted, is reported to have caused a 7-hour breakdown in AT&T's long-distance communication network in the USA in January 1990. The network had been designed to be fault-tolerant and the excessive length of the downtime was presumably due to the unexpected nature of the failure mode.

rate for the 'system failure' mode should not include the failure rates of the alarm circuitry components. Separate calculations can be made to find: (a) the summed failure rates of all components in the system (necessary for calculating how often a repairman will be needed); (b) the failure rate of the alarm system itself (which can be used as a basis for deciding how often the alarm circuitry should be checked for dormant failures).

More complex systems which have redundant subsystems can be analysed using the methods described above. Where necessary, subsystem software failure rates can be added to the corresponding hardware failure rates. Assessment of software failure rates has been described in Chapter 13.

15.5 Choice of Component Failure Rate Data

Broadly, the choice of which data to use will depend upon the use to which the system will be put. For non-military systems, published data from BT's HRD or from Bellcore may be used; where appropriate, in-house data may be used, although such data are usually considered to be confidential so that publication of results may be restricted. For military systems, MIL-HDBK-217 will usually be required with additional data from DEF-STAN 00-41 (Part 3) or NPRD as necessary.

When the reliability engineer can make his own choice of data source, he must use his own engineering judgement; if time and money allow, two or more different data sources may be used and the results compared. The reader is warned that there is a good deal of variability in component failure rate data (see section 5.12.1).

15.6 Calculation of Numerical Values of Reliability Parameters

15.6.1 System Example

Step (d) of section 15.2 is best described by use of an example. It is assumed that a particular failure mode has been identified ('total system failure') and that the corresponding RBD is shown in Figure 15.1. The MUT and MDT are required. The Mission Profile shows that the system will be used in a Ground Benign environment throughout its Design Life. Components having Quality Level 2 (HRD4 data) will be used.

Each of the subsystems shown has been found to have a CFR since the two units in parallel are considered to have a constant failure rate of $2\lambda_b^2/\mu_h$ on the basis that $\mu_h \gg \lambda_b$ (see section 11.11), where λ_b is the unit failure rate and μ_h is the hardware restoration rate (unit b does not contain any software). Subsystem 3 (unit c) is the only one which contains software.

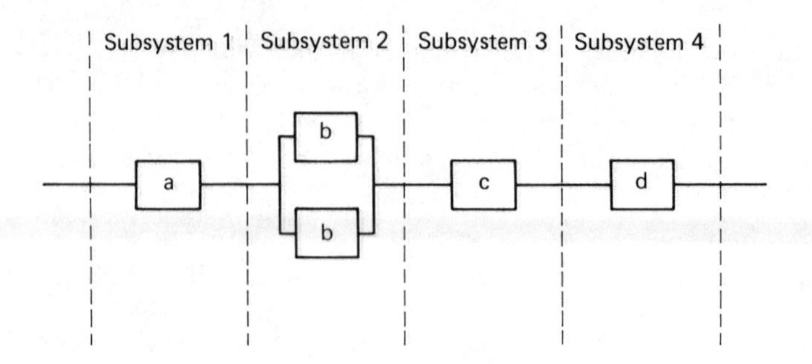

Figure 15.1 RBD corresponding to total system failure

15.6.2 Use of CFR for Software Reliability

Following the methods of Chapter 13, a software failure rate of 30 per cent of the hardware failure rate has been assumed for unit c. Failure rates of the unit have been calculated to be

Unit a - 10 000 fits Unit b - 500 000 fits Unit c - 15 000 fits (hardware only) Unit d - 15 000 fits	HRD4 failure rates used (Quality Level 2)

The hardware restoration rate μ_h has been estimated to be 0.1/hour and there are as many repairmen as faults.

The software restoration rate μ_s is estimated to be 0.5/hour.

Calculation of the total system failure rate is then as shown in Tables 15.1 and 15.2.

The failure rate for the given failure mode is approximately constant and equal to 50 000 fits, which corresponds to an approximate MUT of 20 000 hours (see Exercise **15.2** and Answers). In calculating the approximate MDT, it must be noted that it comprises two parts: that for hardware and that for software.

The hardware MDT (MDT_h) itself has two parts: restoration of two parallel b units (when both have failed) has a mean of $1/2\mu_h$ hours because we have assumed that there are as many repairmen as faults; restoration of any one of the units a, c or d has a mean of $1/\mu_h$ hours because only one repairman will work on a single failure. Following the methods of section 11.8, we should expect the composite MDT_h to be the weighted average of $1/\mu_h$ and $1/2\mu_h$; this is

$$\frac{\lambda_u}{\lambda_u + 2\lambda_b^2/\mu_h} \cdot \frac{1}{\mu_h} + \frac{2\lambda_b^2/\mu_h}{\lambda_u + 2\lambda_b^2/\mu_h} \cdot \frac{1}{2\mu_h} = \frac{\lambda_u + \lambda_b^2/\mu_h}{\lambda_u + 2\lambda_b^2/\mu_h} \cdot \frac{1}{\mu_h}$$

where λ_u is the *hardware* failure rate, excluding that of subsystem 3. The calculation is shown in Table 15.3.

Table 15.1 Unit failure rates

Unit	*Hardware failure rate (fits)*	*Software failure rate (fits)*	*Total failure rate (fits)*
a	10 000	0	10 000 (λ_a)
b	500 000	0	500 000 (λ_b)
c	15 000 (λ_{ch})	5 000 (λ_{cs})	20 000 (λ_c)
d	15 000	0	15 000 (λ_d)

Table 15.2 Calculation of total failure rate (including software) for given failure mode

Sub-system	*Failure rate (symbolic)*	*Failure rate (numerical) (fits)*	*Percentage contribution to total*
1	λ_a	10 000	20
2	$2\lambda_b^2/\mu_h$	5 000	10
3	λ_c	20 000	40
4	λ_d	15 000	30
Totals		50 000	100

Table 15.3 Hardware failure rate, excluding subsystem 2

Subsystem	*Failure rate (symbolic)*	*Failure rate (numerical) (fits)*
1	λ_a	10 000
3	λ_{ch}	15 000
4	λ_d	15 000
Totals	$\lambda_u = \lambda_a + \lambda_{ch} + \lambda_d$	40 000

This is the same formula as that found from an exact analysis (see Exercise **15.1** and Answers) and yields a figure of 9.1 hours for the MDT_h. The MDT_s is $1/\mu_s$, and the composite MDT is the weighted average of MDT_h and MDT_s, namely

$$\frac{\lambda_{hw}}{\lambda_{hw} + \lambda_{sw}} MDT_h + \frac{\lambda_{sw}}{\lambda_{hw} + \lambda_{sw}} MDT_s$$

where the total software failure rate is $\lambda_{sw} = \lambda_{cs}$ and the total hardware failure rate is $\lambda_{hw} = \lambda_u + 2\lambda_b^2/\mu_h$. Substitution of numerical values in the above formula yields

$$MDT = \frac{45\,000}{50\,000} . (9.1) + \frac{5\,000}{50\,000} . (2.0) = 8.4 \text{ hours}$$

Precise calculation of the MTTFF (mean time to *first* failure) shows that this also can be approximated by the MUT found above. We would expect this to be true, since the system is very close to having a CFR. A check should always be made to ensure that the failure rates assumed are compatible with the environmental states shown in the Mission Profile. Changes in the environment which affect the failure rates can be accommodated using the methods of section 6.11.

15.6.3 Use of Software Reliability Functions

If it is assumed that the software reliability can be expressed as a function $R_{sw}(t)$, then the system reliability $R_s(t)$ can be calculated as

$$R_s(t) = R_{hw}(t)\, R_{sw}(t)$$

where $R_{hw}(t)$ is the reliability function for hardware only. In the example above, $R_{hw}(t)$ is approximately $\exp(-\lambda_{hw}t)$ where $\lambda_{hw} = 45\,000$ fits, since λ_c does not now have a 5000 fit contribution for the software failure rate.

Following one of the models suggested in Chapter 13, or by the use of engineering judgement, let the software reliability be assessed at 0.95 for $t = 5000$ hours. That is

$$R_{sw}(5000) = 0.95$$

Now

$$\begin{aligned} R_{hw}(5000) &= \exp(-\lambda_{hw} 5000) \\ &= \exp(-0.225) \\ &= 0.80 \end{aligned}$$

The system reliability at 5000 hours is thus predicted to be $0.80 \times 0.95 = 0.76$.

15.7 Confidence Interval Estimation

Estimation of the Confidence Interval on the MTTF of a system is mathematically complicated and not usually carried out. Some information is given in Johnson (1969).

15.8 Use of Computers for Reliability Calculations

Calculations like the above may be done using a hand calculator, personal computer (PC), minicomputer or mainframe computer. Programs (particularly for PCs) are constantly being introduced or updated and the reader is urged to examine the range of software which may be available. Software is now available which not only holds component failure rate data (for example, from HRD, MIL-HDBK-217 etc. - see sections 5.12 and 5.13) but can also make system reliability predictions. The programs are usually written for use on IBM-compatible personal computers or on Digital Equipment Corporation VAX minicomputers. In the latter case, several workstations are able to operate simultaneously. Facilities are usually available for additional component failure rates to be entered or an in-house library of failure rate data to be accessed. Several companies supply ranges of reliability programs: see section 19.4.

However, these programs should never be regarded as tools which can be used blindly by the inexperienced. A competent engineer with a good basic understanding of reliability theory will save himself much time and effort by using suitable programs; but a user who has no basic understanding is likely to make mistakes which will be compounded by blindly using computer software.

15.9 The Accuracy and Usefulness of Reliability Predictions

The reader who has progressed thus far in the book will not, it is hoped, be surprised to learn that reliability predictions should not be regarded as *accurate* forecasts of what will happen to real systems in the field. Any method of forecasting (from horse-racing results to the weather) is liable to gross error and reliability is no exception, despite the impressive mathematical techniques and the wealth of failure rate data. There are far too many factors outside the control of the reliability engineer (Quality Control during component and system manufacture, design inadequacy, mishandling etc.) for him to present anything more than tentative predictions. It is thus quite unreasonable for him to quote an MTTF accurate to more than 3 significant figures - and 2 would often be more reasonable. When a system is first introduced into service it does not usually have such good reliability as mature equipment (see Chapter 17 on Reliability Growth). In unusual cases (such as space systems), the luxury of improving system reliability with time may not be feasible and every attempt must then be made to get the system 'right first time'.

In cases where systems may be allowed to mature, then reliability prediction obtained by the methods described above should be regarded as predicting the reliability of the mature equipment. The usefulness of reliability predictions is sometimes questioned, since accuracy is not a characteristic which can be claimed for them.

Reliability Predictions should be carried out on any system for the following reasons:

(1) A reliability prediction must of necessity be based on an examination of the system failure modes. This can lead system designers to a better understanding of system behaviour and possibly uncover hazardous failure modes (see section 19.3 on Failure Mode Effect and Criticality Analysis). As a result of the examination, the design of the system may be improved.
(2) It is sometimes required in a System Specification that a reliability prediction should be produced. In any case, system designers and manufacturers should make a prediction of some sort of system reliability, if only to convince themselves that their systems are viable.
(3) A system reliability analysis provides a basis for comparison of (a) competing systems and/or (b) other alternative designs.
(4) Maintenance planning (for example, Integrated Logistics Support) requires reliability predictions.

Exercises

15.1. Draw a Markov diagram for hardware failures only in the system described in section 15.6.1 whose RBD is shown in Figure 15.1. Use the diagram to calculate algebraic expressions for the MDT_h (the Mean Down Time associated with the hardware failures only). Using the failure and restoration rates given in section 15.6.1, calculate the numerical value of the MDT_h.

15.2. Extend the Markov diagram of Exercise **15.1** and prove that the formula given in the text (section 15.6.1) for the MDT is exact. Calculate an exact expression for the MUT and compare it with the approximate value obtained in the text of 20 000 hours.

15.3. Calculate the reliability of the system described in section 15.6.1 at 1000 hours, given that the software reliability is assessed to be 0.90 at this time.

16 Reliable Systems: Specification, Design, Manufacture and Installation

16.1 Introduction

Flowcharts showing how the initial and pre-production design phases can be conducted were given above in Figures 2.1 and 2.2. It was suggested that product objectives should be set for any proposed new system, and it is the purpose of the various design phases to get progressively more confident assessments of whether or not the initial product objectives can be met. If there is extreme difficulty in meeting some product objectives, it is not necessarily the case that the whole project must be abandoned. It may be that they can be modified while still leaving the project viable. The complete specification of a system consists of a number of product objectives relating to Technical Performance, Cost, Quality, Reliability and Maintainability, etc.

'Systems' can cover such a wide variety that only broad recommendations can be given. In any particular case there is no alternative to a very careful analysis of (a) what the buyer expects of the system and (b) how the system is intended to behave (its modes of operation) and how it is likely to fail (modes of failure). There is such a wide range of systems (literally from aerospace to undersea, with everything in between!) which can be considered, that only a broad treatment is possible. Guidelines are set down for anyone wishing to compose the reliability-related parts of a specification. Given that a specification has been written, some methods of system design are discussed and notes on the manufacturing of reliable systems are provided. Reliability aspects of System Installation are considered in section 16.6.

The final parts of the life cycle are field maintenance (see Chapter 14) and disposal (scrap or withdrawal), although disposal is beyond the scope of this book.

16.2 System Specification

16.2.1 Design Life

The Design Life of a system is the length of time that it is designed to survive (with interim repairs if necessary). It should be a realistic assessment of the length of time that the system is expected to be in use, assuming that the scheduled maintenance procedures are carried out. If the Design Life is made longer than necessary, then designers may be forced to use more expensive components than they would otherwise do (because the expensive components have longer useful lives) or make unnecessary provision for the early replacement of some components. If the Design Life is made too short and a decision is subsequently made to extend the system's life-span, then a careful re-appraisal of the components will be necessary and possibly also a costly preventive maintenance programme for the replacement of ageing components which are subject to wear-out.

16.2.2 Mission Profile or Environment

If it is specified, the Mission Profile should give adequate information to the system designer to use for reliability predictions; otherwise, the environments which the system will encounter (and the length of time spent in each one) should be adequately described.

16.2.3 Identification of Failure Modes – General

An important step in composing the reliability section of a system specification is the identification of the hardware and software failure modes which are of importance to the system user (and those important to the system owner if he is not also the user). Once these have been identified, then a reliability requirement can be placed upon each one.

The identification of system failure modes was introduced in section 6.2, where some hardware failure modes of a stereo radio cassette recorder were listed.

A complex system like a telephone exchange has a number of failure modes which include: loss of service to a particular subscriber; loss of service to a particular group of subscribers; failure of the complete exchange; together with failure modes related to severance of a connection, inability to make a call, loss of billing information etc. Identification of possible failure modes should be made by careful analysis of the system's functions and (if possible) by reference to the manner in which earlier similar systems have failed.

For any system, the hardware and software failure modes should be treated separately (wherever this is possible) since their causes are different and alleviation of the effects of failure may be achieved in different ways.

The reliability-related parts of the system specification place an appropriate reliability parameter (MTTF, Availability etc.) against each failure mode. It is important to avoid ambiguity in stating the reliability product objectives for a system. Well-defined failure definitions are essential if a proper reliability study of a system is to be made. In complex systems, there is sometimes difficulty in adequately defining a failure; sometimes it is less ambiguous to use the term 'undesirable event'.

Reliability-related product objectives will be assigned to a system taking into account the use (or uses) to which the system will be put and the Mission Profile which it will have (see Chapter 2).

16.2.4 (Hardware) Reliability Parameters - General

Those parts of the product objectives which relate to hardware reliability parameters could be in the form of any one or a mixture of MTTF, MUT, a required mission reliability ('dependability' in the System Effectiveness equation of Chapter 1), a probability of surviving for a certain time, a steady-state availability, a maximum number of repairs in a given period of time etc.

Unexpected and/or unsafe failure modes can often be identified by the use of Failure Mode, Effect and Criticality Analysis (FMECA), see section 19.3 below. However, the design must usually be in an advanced state before an FMECA can be carried out, and at the specification stage it may not be possible to anticipate all of the significant failure modes.

Some skill is needed in setting product objectives in that a judicious blend of the desirable and the feasible should be aimed at. As an example, a product objective of 100 per cent availability may be desirable, but it is certainly not feasible. But an availability of 99.9 per cent, 99.99 per cent or even 99.999 per cent may be feasible and may constitute a reasonable product objective.

It should be stated how the prediction or measurement of each quoted reliability parameter should be made. For example, it can be stated that a prediction should be made using data from HRD4 (see section 15.2); or that a demonstration test is required (see Chapter 18); or that a measurement using field data should be made after the system has been put into operation. Any one or a combination of these may be called for.

16.2.5 Component Quality

This can be specified or left to the designer's judgement. If it is specified, then some method should be given for the designer to obtain a waiver of the specification so that he can consider the use of non-compliant components. (New components might be available which have not, at the time when design commences, complied with the quality requirements.)

16.2.6 Verification of Compliance of the Design with Specification

If a hardware reliability prediction is called for, then the specification can ask for the manufacturer to provide detailed calculations upon request.

To verify compliance with the software requirements, the specification might call for the manufacturer to disclose the quality control procedures and software tests that he has applied.

To verify compliance with field measurements, it is possible to specify how the field data are to be collected and also to specify a procedure for dealing with relevant and non-relevant failures - in particular with fault-not-found.

16.2.7 Identification of Software Failure Modes

Most systems now contain a certain amount of software and it is desirable that product objectives should be assigned to the software either separately or as part of the total system. As indicated in Chapter 13, however, analysis and prediction of software reliability is much more difficult than for hardware reliability, and identification of how software might fail is consequently a complex problem.

16.2.8 Software Reliability Parameters

Possible requirements that can be placed upon the software include:

(a) Requirements on how the software is developed.
(b) Specification that automatic recovery from a software crash must be provided. Then it is possible to specify a maximum value for the Time to Recovery (or for the Mean Time To Recovery).
(a) A requirement that other defensive measures (apart from automatic recovery) must be taken.

16.2.9 Examples of Reliability-related Product Objectives

Some reliability-related Product Objectives for various systems are shown in Table 16.1.

16.2.10 Other Reliability-related Requirements

(a) Security (see section 16.4.9)
It may be worthwhile to require that:

(1) Data must be secured against loss (as a result of power failure or other causes) and corruption.
(2) Data must be protected against being read and/or corrupted by unauthorised people.

(b) Maintenance (see Chapter 14)
The specification may call for:

(1) provision of alarms;
(2) provision of test equipment;
(3) documentation (such as a Maintenance Manual);
(4) spares provisioning;
(5) training of customer's maintenance personnel.

(c) Safety
The system may be required to comply with safety standards, such as BS 6301: Safety of Equipment that is not part of the Public Network.
(d) Electro-Magnetic Compatibility (EMC)
The system may be required to comply with a standard on EMC.
(e) Contracts based on rewards/penalties
The contract may contain a reliability incentive clause, see section 2.5.2.

Table 16.1 Reliability-related product objectives for various systems

System	*Reliability product objective*	*Notes*
Transatlantic Telephone Cable	Not more than 3 deep water repairs in 25 years	Requirement for TAT-8
Aircraft Automatic Landing System	Probability of a fatal accident due to the use of the complete system must be less than 1 in 10^7	Complete system means both airborne and ground equipment
Telephone Exchange	Not more than 2 hours outage in 40 years	Requirement applies to total exchange failure
Telephone Exchange Subsystem	Mean Total Subsystem Recovery Time not greater than 480 seconds	Requirement on software

16.2.11 Quasi-reliability Requirements

Some requirements in specifications may come under the reliability heading even though they are not strictly matters for a reliability assessment. As an example, a digital communication system may be said to have 'failed' if the Bit Error Ratio (BER) rises above 1×10^{-3} for more than 2 seconds. High BERs are not in general* caused by component failure; they are most likely to occur as a result of external electrical interference. It may be possible to calculate the probability of high BERs using statistical methods, but the unusability of the system is not caused by a 'failure' in reliability terms.

Specifications which refer to 'interruptions of service' can be ambiguous for the reasons given above. It is recommended that specifications should have clearly-differentiated sections, one of which refers to loss of service caused by component failure and others to loss of service caused by other reasons.

Some examples of how loss of service can be caused other than by component failure are: multi-path propagation of microwaves; fading of radio signals; congestion in telecommunication networks.

16.3 The Design of Reliable Systems

16.3.1 Introduction

The initial system design can usually be broken down into a number of subsystems. For each specified failure mode, a series reliability block diagram should be formed and reliability allocations made to each subsystem according to the methods given below in section 16.3.3. It may be clear from the outset that if the product objectives are to be realised, then some redundancy will be necessary. Methods of increasing system reliability have been dealt with in Chapter 7. It should always be borne in mind that methods of increasing system reliability parameters almost always cost money. It may be necessary to make assumptions about the system MDT or other parameters. Improved maintenance (leading to a lower MDT) will always give higher availability, but it is generally more economic to build as much reliability into a system as feasible.

The assumptions made regarding component quality, environment, failure rate source, MDT etc. should always be clearly stated in any analysis in order to avoid misunderstandings.

16.3.2 Source of Failure Rate Data

The system specification should state not only the required reliability parameters but also how they should be verified. In the simplest case the specification gives the failure rate source (such as HRD4) and the designer (or reliability engineer)

*High BERs have been caused in the past by intermittent failures in capacitors. The capacitors broke down for a short period of time (causing a high BER) but the fault was self-healing. This problem proved to be very difficult to trace.

must satisfy himself, and possibly his customer, that the reliability calculations are correct.

If the specification calls for examination of field data or a demonstration test (see Chapter 18) to establish that the reliability requirements have been met, then the designer must (a) use past field data in his calculations or (b) use his best judgement on which data source to use, applying a safety factor to the calculations if he thinks this is necessary. The size of the safety factor must depend upon the risks entailed in not meeting the specification. The risks are not entirely financial and can involve loss of reputation etc.

16.3.3 Reliability Allocation (Apportionment)

For each failure mode, the product objective can be used to give an overall failure rate which can then be broken down into allocations for the various subsystems. In the initial design phase, it is unlikely that any hard-and-fast designs will exist and so it is generally necessary to make some guesses about how reliable the individual subsystems can be made. If the system is repairable (and particularly if it contains redundancy) then estimates must be made of repair times. It is sometimes useful in assessing the failure rate of a subsystem to estimate the number of printed circuit boards which it will contain, and then estimate the average failure rate per printed circuit board.

In a simple series system, failure rates may be initially assigned by using 'complexity factors'. As an example, consider the system of Figure 16.1 which contains four subsystems. The complexity factors are assigned using engineering judgement of the complexity (and hence, to a large extent, the unreliability) of the subsystems.

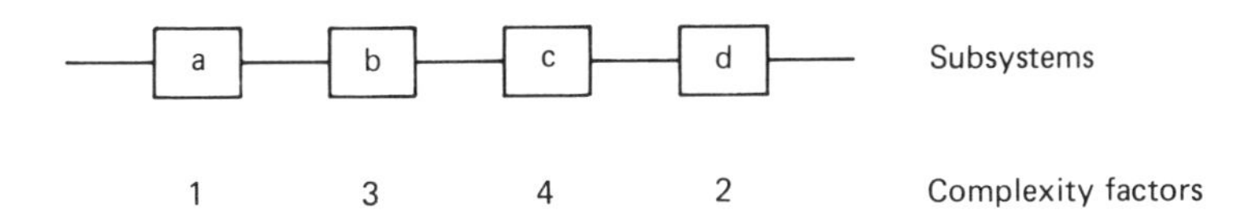

Figure 16.1 A series system with complexity factors assigned to subsystems

The complexity factors are summed (in this case to 10) and the subsystem target failure rates are allocated according to each subsystem's proportion of the total sum. If the system target failure rate in the example were 10 000 fits, then the target failure rates for the subsystems a, b, c and d would be 1000, 3000, 4000 and 2000 fits respectively.

An alternative approach to reliability allocation is for a rough assessment to be made of the failure rate of an average printed circuit board of the size proposed for use in the development. Where data are available, the assessment can be based on previous calculations. Each designer is then asked to assess how many boards he anticipates using in each subsystem, and thus a total failure rate can be calculated and compared with the requirement. Some care is necessary in using this method since not all the components on a printed circuit board will necessarily contribute to the failure rate of the failure mode under consideration. (For example, the alarm circuitry on a board in a telephone exchange may not contribute to the loss of a transmission path through the board.) However, designers can then be persuaded to assess the percentage contribution of each board to the relevant failure mode.

If it is clearly necessary to use redundancy then it should be remembered that a set of parallel redundant units does not have a constant failure rate. However, if each individual unit MTTF is very much greater than the MDT it is generally possible to treat the redundant subsystem as having an approximate failure rate equal to the inverse of the subsystem MUT. For two identical units in parallel, the approximate failure rate is thus $2\lambda^2/\mu$, where λ = unit failure rate and μ = restoration rate. Corresponding approximations can usually be made for subsystems which are m-out-of-n: Good. The allocation process may be iterative and the design, allocation, re-design, re-allocation process will follow the flow-chart of Figure 2.1.

16.3.4 System Design

After reliability allocations have been made to subsystems (and if necessary to units or modules), the proposed designs should be analysed and the predicted reliability performance calculated using methods described in preceding chapters. If the proposed designs do not have the required predicted reliabilities, then methods of improving reliability discussed in Chapter 7 can be applied.

16.3.5 Design for Testability

For ease of maintenance, systems should be designed from the outset to be readily testable. See Chapter 14 on Hardware and Chapter 13 on Software; see also *Guidelines for Assuring Testability* published by the IEE (1988).

16.3.6 Provision of Alarms

The task of the maintenance engineer is made much easier (and the equipment MDT is reduced) if adequate alarms are provided. The provision of alarms should be made an important part of system design from the beginning of the system concept. Human Factors studies have been used to give information on how alarm indications are best provided.

16.3.7 Human Factors

Besides the provision of alarms, it is necessary to consider Human Factors aspects of system signals and controls. Thus it is necessary to consider how best to provide controls which are ergonomically efficient and signals (including alarms) which are comprehensive, explicit and easily understood. These considerations are often classed as part of the study of the man–machine interface.

For portable systems, size, weight and ease of handling are important considerations; ease of installation should be considered for those systems for which it is applicable.

System designers should always remember that it is useless to design a reliable system if it is to be misunderstood or misused by those people who operate or maintain it.

16.3.8 Choice of Components

It was suggested in section 1.6 that an optimum system reliability is attainable. This may not be easy to achieve in practice. For high-reliability systems, the cost of failure in terms of money and/or loss of life may be so high that expensive, highly-reliable components are obviously required. However for other systems, component quality must be chosen in accord with the cost and quality of the system under development.

Some care must be taken if limited-life components are selected for a design. If the service life of the equipment is not much less than the wear-out life of a component, then provision (via the maintenance instructions) must be made for replacement of the component well before wear-out causes the failure rate to rise sharply. Even if the service life is less than the wear-out life of some components, it is still good practice to list known limited-life components in the maintenance manual. Then if the service life of the equipment is subsequently extended, provision is easily made for the limited-life components to be replaced.

16.3.9 The Design of Software

See Chapter 13 for a general discussion of Software Reliability.

It is sometimes possible to design the software so that automatic recovery from a software 'crash' is possible. Such measures make the effect of crashes less severe (or perhaps even unnoticed by the user) and are generally well worthwhile.

16.4 Techniques in the Design of Reliable Systems

16.4.1 Redundancy

When redundancy is used, it is important to eliminate as many common modes of failure as possible. As an example, consider that a very secure connection is needed between a factory site and a public telephone exchange. It may be decided to use two separate telephone cables between the site and the exchange. It would be a mistake to put the two cables in the same duct, since a mechanical digger would sever not one but both cables simultaneously. For high reliability, the two cables should follow completely different routes and, if possible, be connected to two separate public telephone exchanges.

16.4.2 Majority Voting

In some high-reliability systems it may be that a one-out-of-two redundant system may not be sufficiently reliable because a failure in one of the parallel units may not be self-announcing. This can happen, for example, in an aircraft radio-altimeter system. If two independent altimeters are used then their outputs can be compared and, so long as there is sufficiently close agreement, the outputs can be judged to be correct. But if the two outputs disagree, it may not be easy to design a system which can always (or nearly always) correctly decide which output should be ignored.

In these circumstances it is often desirable to use a majority-voting system in which the outputs from three separate altimeters are compared. If there is agreement between all three outputs there is no problem. If one output disagrees with the other two, then the correct output is taken to be that of the majority. If all three outputs disagree, then the system declares itself to have failed.

16.4.3 Diversity

Diversity may be regarded as a form of standby redundancy, although it is not primarily provided to ameliorate the effects of failure. In radio systems, space and frequency diversity are sometimes used to increase the reliability of a link by using two different receiving antennas (space diversity) or two different frequencies (frequency diversity).

The essential factor in diversity reception is that two different signal strengths are compared and the channel with the smaller signal strength is rejected. The process of comparison is on-going, and at a later time the previously rejected channel may prove to have the higher signal strength and then become the accepted channel. In standby redundancy, the changeover from one system to another is normally caused by system failure.

16.4.4 Graceful Degradation

In designing reliable systems it is sometimes useful to employ the concept of graceful degradation. If a system degrades gracefully, then it is designed in such a way that successive failures cause the system to be degraded, but not to fail completely. An example of graceful degradation can be seen in a telephone exchange in which a single failure may reduce the traffic-handling ability of the exchange (or reduce service to a group of subscribers), but is very unlikely to cause the whole exchange to fail. Subsequent failures are again likely to reduce the traffic-handling capacity (or subscriber service) still further, but again are unlikely to cause total failure. Medium-sized public telephone exchanges are designed to have an MUT (for total failure) in the region of 40 years.

16.4.5 Dormant Failures

Systems may continue to operate even though they contain failures. Even series systems may be liable to failures of this type. (For example, a de-coupling capacitor may go open-circuit without necessarily degrading the system so much that it fails.) Systems containing redundancy in particular may contain failures which are not self-announcing. Such failures are called dormant failures and may occur even in systems which contain self-diagnostics and alarms. It is usually important, particularly in the later stages of product development, to search the design for possible dormant failures. So long as they have been identified, then checks for them can be made as part of the preventive maintenance schedule. Failure Mode, Effect and Criticality Analysis (FMECA) is particularly useful for finding dormant failure modes.

16.4.6 Failure Mode, Effect and Criticality Analysis/Fault Tree Analysis

These analytical methods are described briefly in sections 19.3 and 19.2.

16.4.7 Integrity

In some systems (such as the aircraft radio altimeter described in section 15.4.2) it may be important that a system should declare itself to have failed rather than give a false reading caused by failure. This characteristic of a system is called integrity. One definition is 'that quality which relates to the trust which can be placed in the correctness of the information supplied by the system'.

16.4.8 Safety

Safety is a vital element of any system and an analysis of the possible failure modes should always be made to check that none of the foreseeable failure modes will cause the system to be needlessly unsafe. In systems where failure may be particularly catastrophic (such as nuclear power stations) it is important to make the system as fail-safe as possible. It is suggested, however, that a system can never be completely fail-safe since there will always be some chain of events, however improbable, which will cause catastrophic failure. System design should be such that the probability of catastrophic failure is acceptably low.

In the analysis of systems with particular regard to safety, Fault Tree Analysis (FTA) and Failure Mode, Effect and Criticality Analysis (FMECA) are useful tools (see Chapter 19).

16.4.9 Security

'Security' is sometimes used synonomously with 'reliability'. This usage is to be deprecated in engineering literature since the two words should have very different meanings in a technical context. A useful definition of security is 'the condition that results when measures are taken that protect information, personnel, systems, components and equipment from unauthorised persons, acts or influences' (Weik (1983)).

Thus it is possible for a highly reliable system to be very insecure (for example, a reliable radio system which cannot prevent unauthorised people from listening in), and for a very unreliable system to be very secure (for example, an unreliable data transmission system which has an unbreakable ciphering algorithm).

16.4.10 Error Detection and Correction

Error Detection and Correction techniques are used to protect the data transmitted in digital communication systems. They are not fault tolerance techniques in the reliability sense since the digital errors are most likely to be caused by a poor signal/noise ratio rather than by component failure. (See section 16.2.11 and footnote.)

16.4.11 Design for Simultaneous Wear-out

It would appear to be a promising idea for a designer to produce a system (particularly one built chiefly from mechanical components) in which each component had exactly the same wear-out time. If this could be achieved, the system would survive until such time as all the components simultaneously wore out and the system suffered total collapse. The idea is not new and is described in the poem *The Deacon's Masterpiece* by Oliver Wendell Holmes. The 'masterpiece' is a horse-drawn cart which survives for 100 years and then suffers complete collapse. Such designs do not seem to be practicable. A car may have minor components (such as batteries and tyres) replaced several times during its lifetime, and wear-out occurs when a major component (such as chassis or engine) becomes uneconomical to repair.

16.5 The Manufacture of Reliable Systems

16.5.1 Component Procurement

It is obviously useless to spend much time and money in trying to design a reliable product if this work is to be vitiated by the use of inferior components during the manufacturing process. It is the function of Quality Control to ensure that the components used in manufacturing a system (a) are those specified by the designer and (b) conform to their specifications.

16.5.2 System Manufacture

During system manufacture it is necessary to maintain high Quality Control standards on the production line. For example, a badly-adjusted soldering bath could cause the subsequent failure of many systems in the field. The introduction of Computer Integrated Manufacturing (CIM) presents an opportunity for improved Quality Control procedures. In the manufacture of printed circuit boards, a Detailed Assembly Process Planning System (DAPPS) can be installed which keeps a history of each board as it passes through the whole manufacturing stage.

In addition to great emphasis on Quality Control, manufacturers of highly-reliable systems used in an aerospace or undersea environment use specially-constructed clean rooms for equipment assembly. It is also necessary that highly-trained staff should be used for the manufacturing and inspection processes.

16.5.3 System Testing

The manufacturer may wish to carry out his own tests in addition to the Demonstration Tests described in Chapter 18. Statistical analysis of failure data follows the methods of Chapters 3 and 4.

16.5.4 System Burn-in

It is usual for complete equipments (for example, television receivers) to be given some burn-in before they are released for sale. In this way, equipments which might produce early failures in the field tend to be weeded out. Even if the separate components have already been burnt-in, the finished equipment may have inherent defects (for example, in the soldering of the connections) which might cause early failure and system burn-in will help to remove these. Typically for electronic systems, burn-in could be of 50–150 hours duration and at 70°C.

16.6 Installation

This is a key area for reliability but its importance is often overlooked. If improperly installed, an otherwise satisfactory system (from the point of view of design and manufacture) will be unreliable. It is thus important that the installation team should be well-trained and aware of the importance of its work. Installation can be a fairly simple process: for example, installation of a TV receiver will involve connection of mains and antenna plugs, tuning the required channels and picture adjustment. On the other hand, installation of a large system like a telephone exchange involves extensive work which is here taken to include commissioning and testing.

17 Reliability Growth

17.1 Introduction

Except for high-reliability systems like spacecraft or undersea cables, it would be unreasonable to expect the earliest-made systems sold to a customer to show their full potential reliability. Usually there are teething problems in both manufacture and maintenance of new systems, and these cause the reliability to be lower than that predicted.

17.2 Duane's Plotting Method

Duane discovered that a plot of the logarithm of the Cumulative MTTF (that is, cumulative hours/total number of failures) against the logarithm of cumulative hours is often a straight line (see Duane (1964)). The fit is generally poor at the earliest times of failure and it is usual to discount a short period at the beginning to allow for infant failures and other early failure causes.

The fitting of a straight line is a purely empirical phenomenon, but good results have been obtained by a number of researchers. The slope of the straight line is claimed to be a measure of how much attention is given to reliability improvement. A high slope (about 0.6) is associated with a strong reliability improvement programme in which all failures are carefully analysed and improvements are implemented as a result of the analysis. A low slope (about 0.1) is associated with little or no attempt to improve system reliability.

The usual form of a Duane plot is shown in Figure 17.1 which is taken from Mead (1977); the slope is measured to be 0.4. In cases where the data do not fit a Duane plot, it may be useful to plot the data in a CUSUM form.

17.3 Comments

The advantages of using Duane's plotting method are:

(a) Early indications are given of the efficiency of the reliability improvement methods.

(b) Confidence may be generated that a target MTTF may be achieved.
(c) Some customers are willing to accept equipment on the basis that reliability growth has been demonstrated and that a target MTTF is attainable. This cuts the cost of demonstrating the reliability of the final product and saves the time which would be taken up in the demonstration tests.

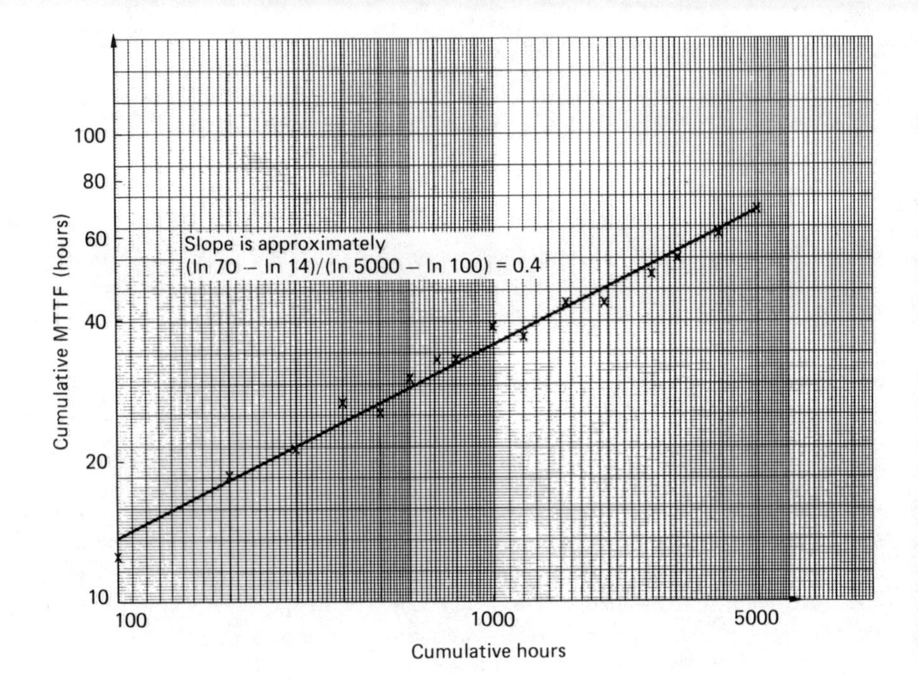

Figure 17.1 A Duane plot showing reliability growth (early data omitted)

The disadavantages of Duane's method are:

(a) There is always a danger in extrapolating early results into the future.
(b) Duane's method has been shown to work well with hardware failures, but may not be so useful for predicting MTTFs when both hardware and software failures are taken into consideration.

18 Reliability Demonstration

18.1 Introduction

Probably the most convincing demonstration of reliability is the systematic production of items (systems or devices) which are recognised by their users to be highly reliable. Rolls Royce cars have an enviable reputation which has been established over a number of years. The reliability of a well-established product may be demonstrated by analysis of past failure data. The reliability of new products, however, must either be accepted on trust or demonstrated by the manufacturer. Section 18.2 briefly surveys how system failure data may be analysed. Section 18.3 onwards is concerned with Demonstration testing.

18.2 Analysis of System Field Failure Data

System failure data obtained from the field can be analysed by the same methods as those described in Chapters 3 and 4 for components. In particular, Weibull plotting of data can be used to establish whether or not the system can be considered to have a CFR. If it does, then point estimates and confidence intervals can be calculated for systems, just as they were for components. If the system does not have a CFR then an attempt should be made to fit a Weibull or some other distribution to the data - see section 4.10.

18.3 Demonstration Testing

At the production or pre-production stage of development of a component or system, it is usually necessary to demonstrate that its reliability is at least as high as some minimum required level. The remainder of this chapter is devoted mainly to hardware demonstration testing, since well-tried methods have been developed over many years. However, most modern electrical systems contain at least some software, and it may not be desirable to differentiate between hardware and software failures. Thus, although the demonstration tests described in this chapter are designed chiefly for hardware testing, they can usually be adapted to deal with systems containing both hardware and software.

Demonstration tests may be needed for components, modules, subsystems or systems; in this chapter, the word 'device' will be used to mean any one of these.

18.4 Sampling from a Population

In attempting to demonstrate the reliability of a device, it is usually possible to take only a small part of the total statistical population for testing. (Exceptions may occur where only a small number of systems will be built, for example, as in the case of systems designed for use in spacecraft.) Every attempt should be made by way of Quality Control to ensure that the total output of devices is statistically homogeneous. The sample taken for testing should then be chosen in a suitably random manner (for example, by the use of random number tables) so as to remove any possible bias from the sample. (For example, it is undesirable that the devices selected for testing should all be taken from the same production batch.) But even after taking all possible precautions, there is still risk for both the producer and consumer that the sample may not be truly representative of the total population.

In designing a demonstration test the producer and the consumer (customer) should try to agree on:

(a) what the test is trying to demonstrate (value of MTTF, proportion of population which is satisfactory etc.);
(b) the risks which both producer and consumer are willing to take.

18.5 Operating Characteristics: Binomial Test

Consider that a sample of size n is drawn from a large population. Let it be possible to characterise each member of the sample as 'good' or 'bad', and that the population is so large that the probability of a member of the sample being good or bad does not vary as the members are drawn successively from the population. (This assumption would not be true if the population size were not very much greater than the sample size. In that case, the development given below would have to use the hypergeometric rather than the binomial distribution.)

Let the probabilities that a sample member is 'good' be q, and that the sample member is 'bad' be p. Then

$$p + q = 1$$

and the probabilities that the sample of n contains the different possible mixes of good and bad members are obtained by the binomial expansion of $(p + q)^n$. In this expansion, the term p^n will represent the probability that all n members are bad; the next term, $np^{n-1}q$ will represent the probability that $n - 1$ are bad and 1 good; and so on until q^n represents the probability that all the members are good.

If an acceptance criterion is agreed between producer and consumer, this will be of the form: 'the test is passed if c or fewer defective items are found in a sample with n members'. The risks involved are shown in an operating characteristic curve like that of Figure 18.1. This is derived by finding the probability of acceptance

$$(q^n + {}^nC_1 q^{n-1}p + \ldots + {}^nC_c q^{n-c}p^c)$$

where ${}^nC_i$ is the number of ways of choosing i items from n and equals

$$\frac{n!}{(n-i)!\,i!}$$

for each value of p. When $p = 0$ (the whole population is good) the probability of passing the test is 1, while when $p = 1$ (the whole population is bad) the probability is 0.

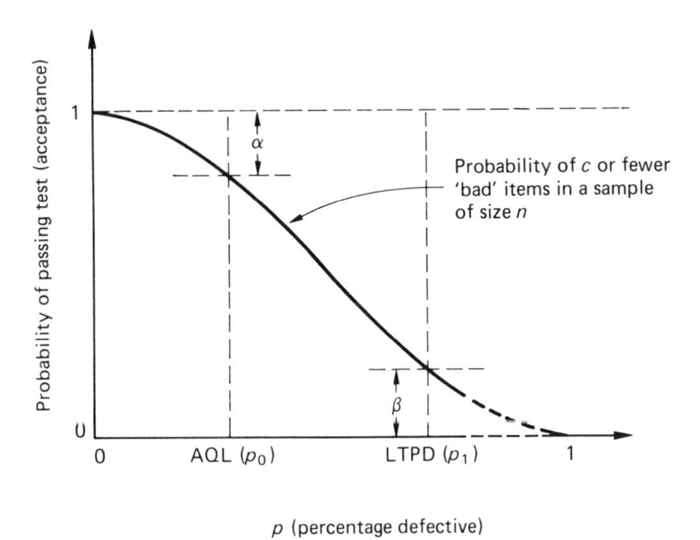

Figure 18.1 Operating characteristic curve

Particular risks are calculated at two values of p: the Acceptable Quality Level (AQL) and the Lot Tolerance Percentage Defective (LTPD) as shown in Figure 18.1. (LTPD is often known in the USA as RQL - Reject Quality Level.) The values of p corresponding to the AQL and LTPD will be called p_0 and p_1 respectively.

The risk α shown in Figure 18.1 is called the Producer's Risk: this is the risk the producer runs that if his product has a proportion of bad items as low as p_0, then the test will not be passed. (The ordinate in the graph of Figure 18.1 represents the probability of acceptance so that the distance between the ordinate and the horizontal broken line represents the probability of rejection.) The risk β is called the Consumer's Risk and represents the risk the consumer runs that if the proportion of bad items is as high as p_1, then a sample of size n may still be accepted.

The ratio p_1/p_0 is called the Discrimination Ratio; as the name implies, it is a measure of the range from p_0 to p_1.

There are six parameters which are generally of interest in a characteristic curve: n, c, α, β, p_0 and p_1. A particular test is defined by four parameters, say n, c, α and β. If a test is required with defined values of α, β, p_0 and p_1, it may be necessary to choose values of n and c which represent compromise values, because n and c must both be integers and it is likely that only approximate values of α and β are achievable.

A nomograph (Larsen Chart) is available for the derivation of an acceptance test (see Larsen (1965) and Larsen (1966)).

An example of an operating characteristic curve for $n = 50$, $c = 1$ is shown in Figure 18.2.

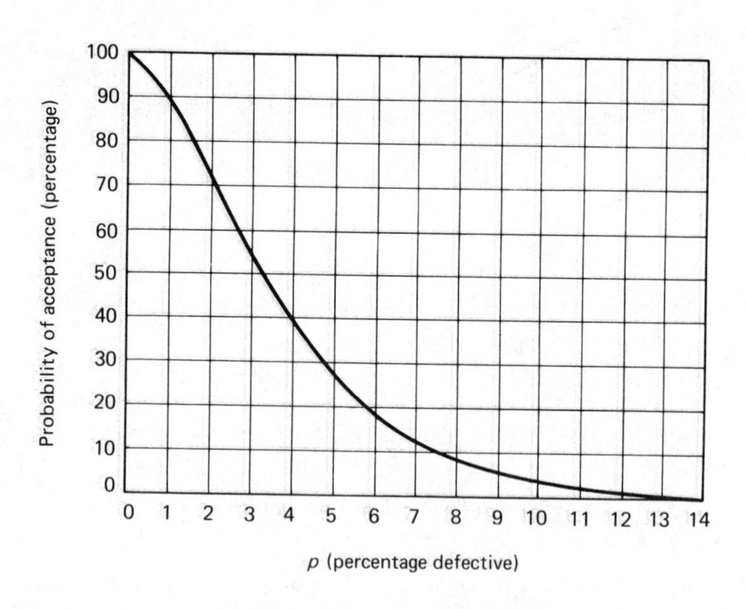

Figure 18.2 An operating characteristic curve for $n = 50$, $c = 1$

If $\alpha = \beta = 0.10$ (10 per cent), then $p_0 = 0.0107$ (approx.) is the solution of the equation

$$(1 - p_0)^{50} + 50(1 - p_0)^{49} p_0 = 1 - \alpha = 0.90$$

and $p_1 = 0.0756$ (approx.) is the solution of the equation

$$(1 - p_1)^{50} + 50(1 - p_1)^{49} p_1 = \beta = 0.10$$

It should be noted that so far no reference has been made to the actual proportion of 'bad' items which the producer is manufacturing. This is called the Process Average Quality (PAQ). The producer will normally try to keep his PAQ near to the AQL, because then he has a high probability of passing the test. (Failure of a test usually has some financial penalty for the producer in that he must do rework on his sample or lose time by submitting another sample for a test.) Considerations of where the PAQ should lie with relation to the operating characteristic curve are discussed in Evans (1981).

18.6 Operating Characteristic Curve for a Demonstration Test: Constant Failure Rate Case

An operating characteristic curve may be constructed using the results of section 3.16. A plot of Pr(0) against λT (or equivalently T/θ) gives the probability that no failure will occur if a device with constant failure rate λ is tested over a total of λT device hours.

Similarly, a plot of Pr(0) + Pr(1) + . . . + Pr(c) gives the probability of c or fewer failures. In the constant failure rate case considered here:

$$\Pr(0) + \Pr(1) + \ldots + \Pr(c) = \exp(-\lambda T) + \frac{\lambda T \exp(-\lambda T)}{1!} + \ldots + \frac{(\lambda T)^c \exp(-\lambda T)}{c!}$$

Curves for $c = 0$, 1 and 2 are plotted in Figure 18.3, which is similar to Figure 18.1 except that the abscissa (λT) ranges from zero to infinity instead of from zero to unity.

It is often more useful to use a logarithmic scale for the horizontal and vertical axes: such a plot is shown in Figure 18.4.

For a test in which T_c device hours are generated, the risks α and β corresponding to failure rates of λ_0 (AQL) and λ_1 (LTPD) with $c = 2$ are shown in Figure 18.3. The Discrimination Ratio (or, in reliability theory, the Reliability Design Index) is $\lambda_1 T_c / \lambda_0 T_c$ which reduces to λ_1/λ_0 or θ_0/θ_1 where the θs are the corresponding MTTFs.

18.7 Operating Characteristic Curve for Demonstration Test: Non-constant Failure Rate Case

If the failure density function of the device under test is a known mathematical function, then it may be possible to develop the equation of the operating characteristic curve by methods similar to those of section 18.4 by convolving the known density function.

If analytical methods are impossible, either because the convolution cannot be carried out analytically, or because the failure density function is not known in an analytical form, then numerical methods (again using the methods of section 18.4) may be used.

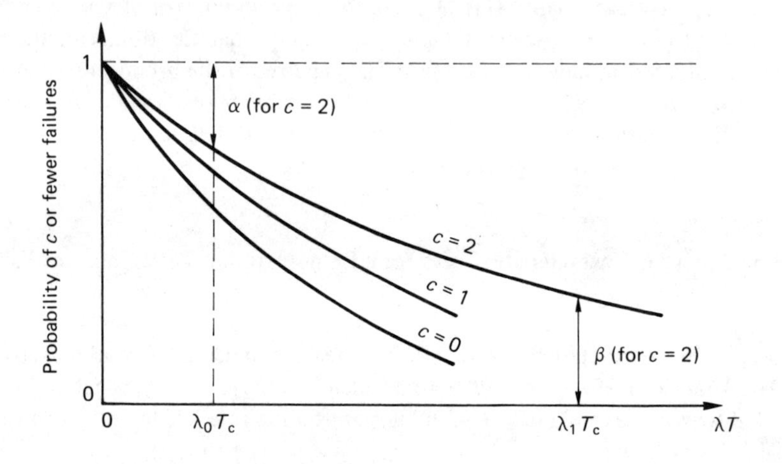

Figure 18.3 Probability of c or fewer failures as a function of λT

Care must be taken in this case in assembling the data since the device under test does not have the memoryless property of the constant failure rate device and so it is no longer permissible to lump the device hours together.

18.8 Problems in Demonstration Testing

Demonstration tests of the type described in sections 18.5 and 18.6 suffer from some disadvantages since to demonstrate the required MTTF a large number of device-hours on test may have to be generated. This is expensive and time-consuming, since usually the testing may only commence near the end of the development phase. The number of device-hours and failures needed to generate an upper 60 per cent confidence limit on the failure rate is shown in Table 18.1. Taking into consideration the fact that the failure rate (under normal operating conditions) for an electrolytic solid tantalum capacitor is in the range 1–5 fits and for an MOS digital IC is in the range 50–300 fits, it can be seen that very large number of component-hours on test must be generated in order to give confidence to a failure rate prediction for a component.

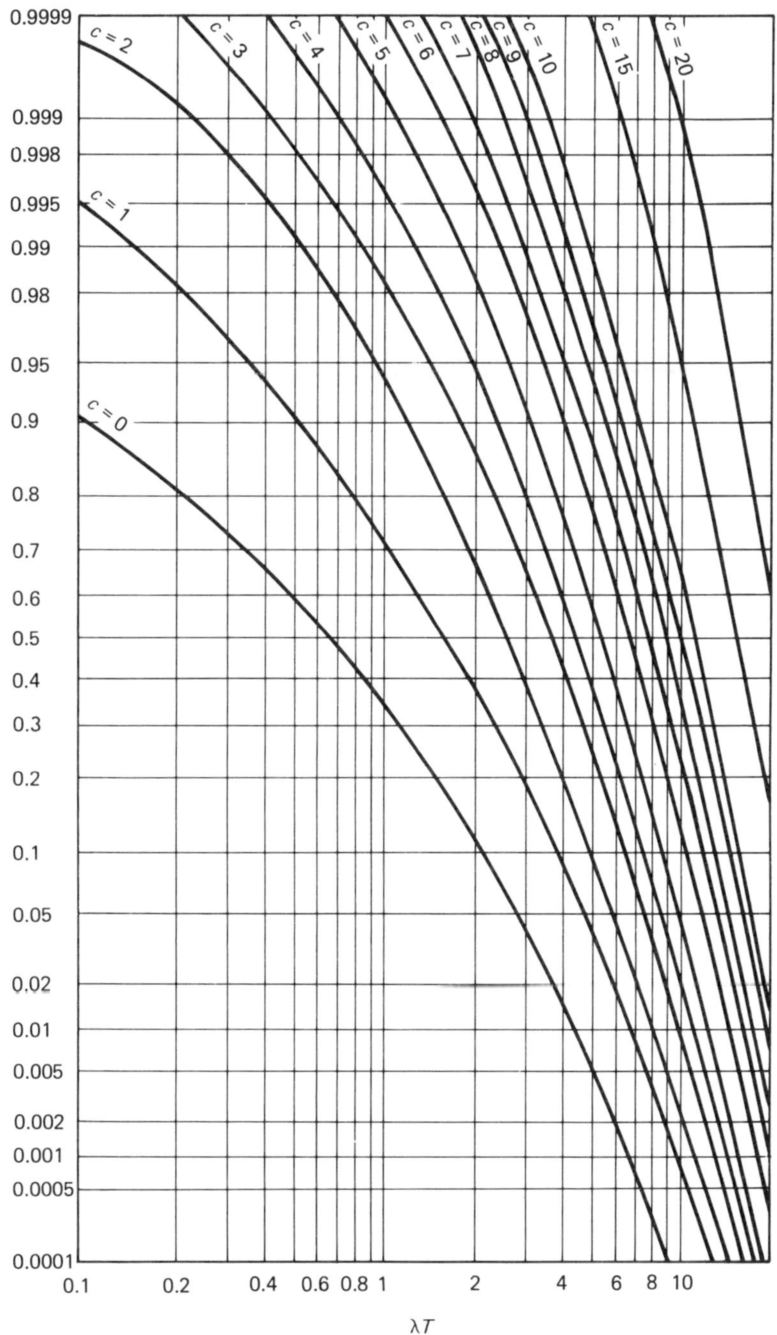

Figure 18.4 Probability of c or fewer failures as a function of λT

Table 18.1 Device-hours which must be generated for failure rate measurement

Device-hours generated	*Number of failures*	*Estimated failure rate (fits)* Maximum likelihood estimate	*Estimated failure rate (fits)* Upper 60 per cent single-sided limit
100 000	1	10 000	20 000
	3	30 000	41 800
1 000 000	1	1 000	2 020
	3	3 000	4 180
10 000 000	1	100	202
	3	300	418
	10	1 000	1 150
100 000 000	1	10	20
	3	30	42
	10	300	321
	100	1 000	1 030
1 000 000 000	3	3	4
	10	10	12
	30	30	32
	100	100	103

To overcome these disadvantages, it is possible to:

(a) use accelerated testing (see below);
(b) use a graphical prediction method (see below);
(c) use sequential test methods (see below);
(d) rely on a demonstration of reliability growth (see Chapter 17).

18.9 Accelerated Testing

Since a large number of device-hours of testing is usually necessary under normal ambient conditions, accelerated testing is often employed. In an accelerated test, the device is subjected to higher stress levels (particularly temperature) than it would normally withstand. The higher stresses cause the failure rate to increase, and the required number of failures for adequate prediction will occur in a shorter time than under normal testing.

The failure rates under normal ambient conditions are then found by extrapolation. If the activation energy of the failure process is known or can be estimated, then Arrhenius's equation (see Chapter 5) can be used for extrapolation from high temperature to normal ambient. Sometimes a combination of high temperature and high humidity is used in accelerated testing. The effect of humidity on failure rates is not so well understood as is the effect of temperature.

Great care must be taken with accelerated testing since extrapolation of the failure rate back from high stresses to normal stresses assumes not only that the same failure modes apply but also that their relative proportions remain unchanged. It can happen that different failure modes (or different relative proportions of them) appear at the high stress levels from those at normal stress levels, and the extrapolation is then invalid.

18.10 A Graphical Prediction Method

Some customers are willing to accept a reliability prediction based upon Weibull plotting. A small sample of devices is put on test and the first few failures are plotted on Weibull Probability Paper. By extrapolation backwards, a prediction of a low value (say 1 per cent) of time to failure for the population may be found. So long as the lower confidence limit of this low percentage value exceeds a certain agreed value, then the device is deemed to have passed the test.

18.11 Sequential Reliability Demonstration Tests - Untruncated Case

In order to reduce the time taken to reach a decision, Wald (1947) introduced the idea of a Sequential Probability Ratio Test (SPRT); such tests can be used in a wide range of disciplines but are particularly useful in the field of reliability. The tests apply only to constant failure rate devices, and for that reason it is usual to apply some burn-in before the test commences. (Infant failures produce low times-to-failure so that if infant failure devices are included in the test batch, the results will be unduly pessimistic.)

In a sequential test, the cumulative number of failures (F_c) is plotted against the cumulative test time (T_c). Such a plot is shown in Figure 18.5 in which the first failure occurred after a total of t_1 cumulative test hours, the second after t_2 hours and the third after t_3 hours. In an actual test, the data are plotted in this way until the plot crosses one of the two boundary lines, when the test is stopped and the appropriate decision is taken. The equations which define the accept and reject boundaries are obtained from the parameters α, β, θ_0 and θ_1 (or α, β and θ_0/θ_1) which are chosen prior to the test and the meaning of which has been explained above.

Wald was able to show that so long as both α and β are small (say less than 0.20) then the boundary equations are:

$F_c = mT_c + (\ln A)/G$ for the upper (reject) boundary

and

$F_c = mT_c + (\ln B)/G$ for the lower (accept) boundary

where

$$m = \frac{(1/\theta_1 - 1/\theta_0)}{\ln(\theta_0/\theta_1)} \qquad G = \ln(\theta_0/\theta_1)$$

$$A \approx (1-\beta)/\alpha \qquad B \approx \beta/(1-\alpha)$$

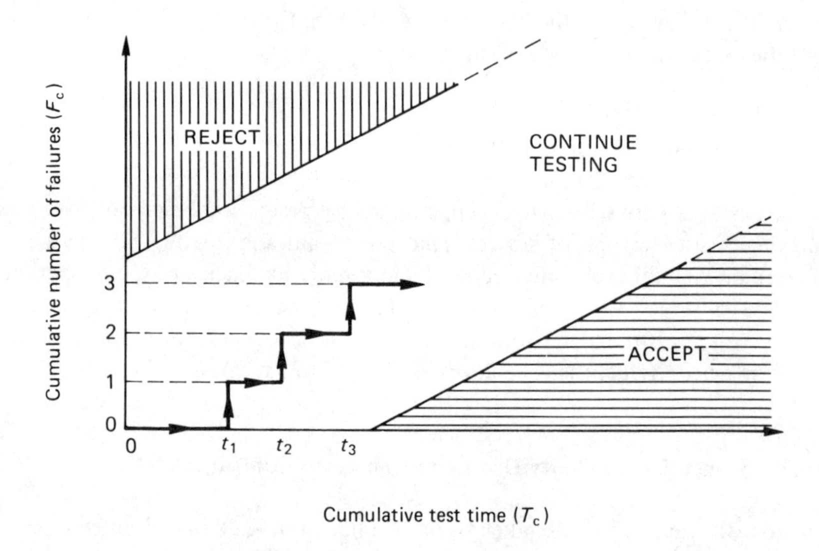

Figure 18.5 An untruncated sequential test

The operating characteristic (o-c) curve for a sequential test can be calculated: see Mann *et al.* (1974).

Care must be taken in sequential testing (as in any form of demonstration test) that non-relevant failures are excluded from the test. Environmental conditions (possibly including environmental cycling) for the test must be agreed between producer and consumer.

18.12 Sequential Reliability Demonstration Tests - Truncated Case

A disadvantage of the sequential test illustrated in Figure 18.5 is that, although the average time to reach a decision is much less than in the fixed-time test, there is a (small) possibility that the test could go on for a very long time before a decision is reached. For this reason, truncated sequential tests have been introduced. An example is shown in Figure 18.6 in which the test is truncated either at F_T failures or at cumulative test time T_T. Unfortunately, the effect of truncation is slightly to change the equations for the boundary lines which are not as easily derived as in the untruncated case.

Examples of truncated test plans are quoted in MIL-STD-781C *Reliability Design Qualification and Production Acceptance Tests: Exponential Distribution*, produced by the US Department of Defense:* environmental conditions are also quoted.

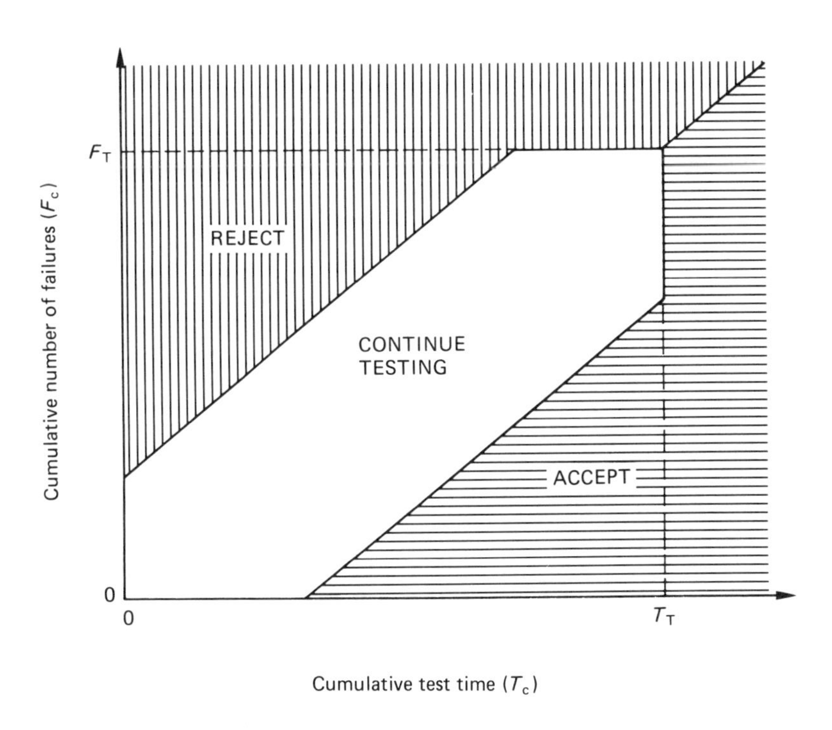

Figure 18.6 A truncated sequential test

18.13 The Selection of Values for θ_0 and θ_1

The selection of values for θ_0 and θ_1 in a Demonstration test can represent a considerable problem, because it tests a reliability prediction against real results. The predicted system reliability may be inaccurate for use in the Demonstration test for many reasons, but chiefly because: (a) the failure rate data used for the prediction may be excessively optimistic or pessimistic; (b) early-produced systems may have poor initial reliability which may be improved by a Reliability Growth

*In this standard, the abscissae of the test plans are given in multiples of θ_1 (the 'specified MTBF').

Programme. The producer may thus suggest that θ_0 and θ_1 should be lowered to take account of these effects; the consumer will naturally try to oppose the suggestion. The selection of values θ_0 and θ_1 which are agreeable to both producer and consumer is a difficult problem calling for good engineering judgement (which should not be regarded as a euphemism for guesswork!).

19 Some Analytical Methods and Computer Software

19.1 Introduction

In this chapter, the topics of Fault Tree Analyses (FTA) and Failure Mode, Effect and Criticality Analysis (FMECA) are covered briefly. References to more detailed descriptions have been added where necessary. In addition, section 19.4 lists some of the computer software which is currently available commercially.

19.2 Fault Tree Analysis

Fault Tree Analysis (FTA) is a logical procedure for finding the causes of faults or 'undesirable events'. The most commonly-used symbols for assembling a fault tree are shown in Figure 19.1. In its simplest form, an FTA is a graphical form of a logical expression. As an example, consider the system shown in Figure 9.1. The minimal path set for this system has been found in section 9.5. Corresponding to a minimal path set, it is possible to define a minimal cut set (see Bansal *et al.* (1982)):

> A cut set is a set of branches which when cut do not leave any path from input node to output node.
> A minimal cut set is a set of branches which satisfies the above property but no subset of which has this property.

The definitions are rather unwieldy, but in our example $\bar{A}\ \bar{B}\ \bar{C}$ is a cut set.* This, however, represents overkill; it is only necessary for units a and b to fail for the system to fail. Thus $\bar{A}\ \bar{B}$ is a minimal cut set since if we remove either $\bar{A}$ or $\bar{B}$, the result is not a cut set.

It is easy to see that for the system of Figure 9.1, the set of minimal cut sets is

$$\bar{A}\ \bar{B} + \bar{B}\ \bar{C} + \bar{C}\ \bar{D}$$

*The capital letters are used here to represent events. A is the event that unit a is 'up', $\bar{A}$ that it is 'down', with a similar nomenclature for the other units.

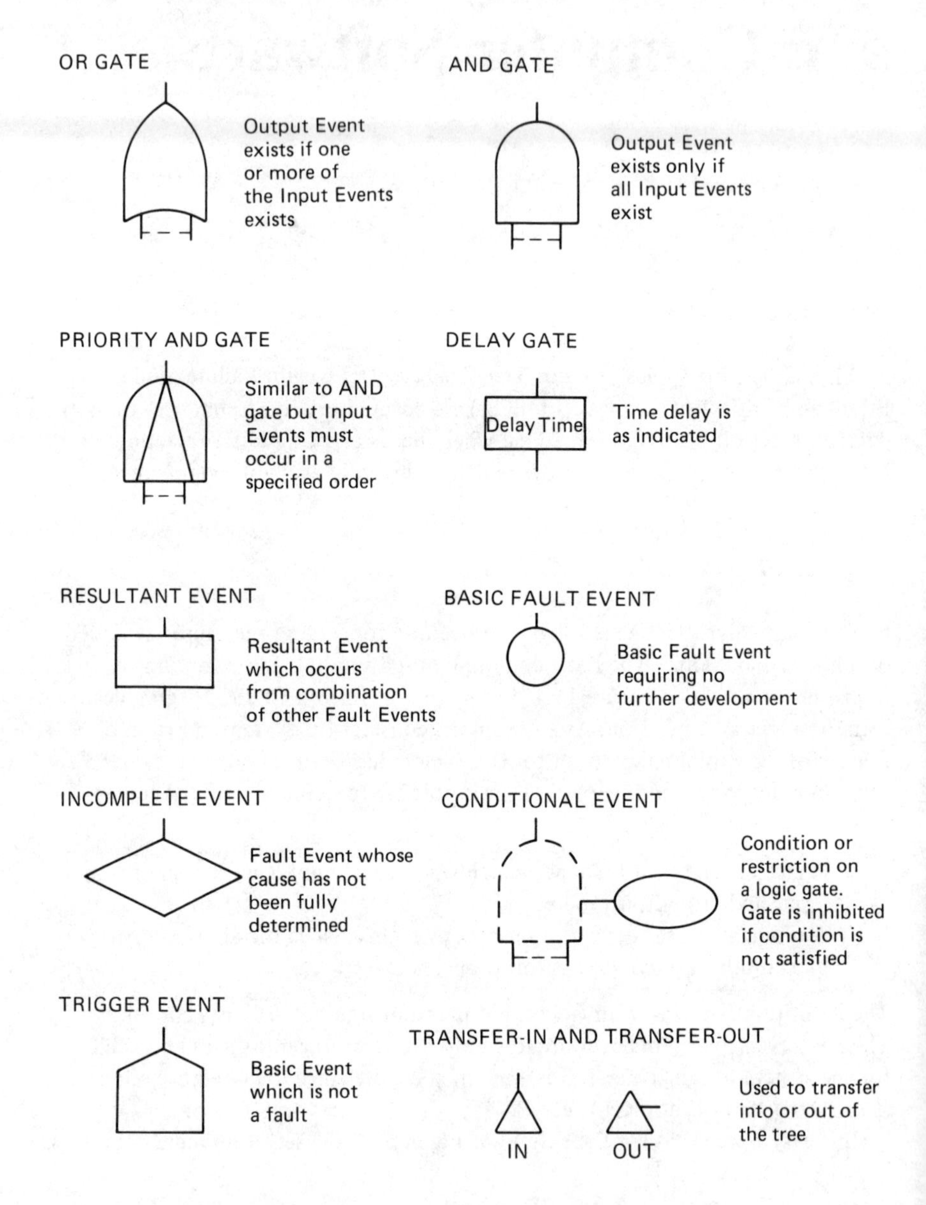

Figure 19.1 Logic symbols used in fault trees

Using the symbols of Figure 19.1 the Fault Tree for the system of Figure 9.1 is shown in Figure 19.2.

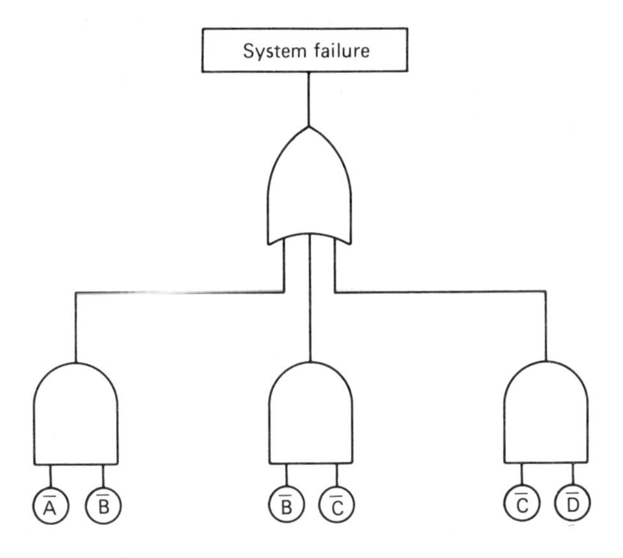

Figure 19.2 Fault tree for the system shown in Figure 9.1

The example given above is very simple; fault tree analysis is usually more complex and not so straightforward as the above example might suggest. It is particularly useful for the analysis of systems whose failure can result in a condition which is hazardous to life. For further information, see: Bansal *et al.* (1982) on path sets and cut sets; O'Connor (1985) or Lewis (1987) on FTA.

19.3 Failure Mode, Effect and Criticality Analysis (FMECA)

As the name suggests, this is an analysis of the ways in which a system can fail (the failure modes), the effects of such failures and a measure of the seriousness of the mode (the criticality). The analysis is usually presented in a tabular form; a very simple example is shown in Table 19.1 in which an FMECA is presented for the stereo cassette radio described in sections 6.2 and 6.4.

The process of analysis can be taken through subsystem, unit, module and even down to component level. As an example of how component modes of failure can affect the criticality of the failure, we consider the mains power switch on the stereo cassette radio player. If the switch fails into open circuit the whole radio is unusable and the failure is of criticality 1 (the highest criticality). If the switch fails into short-circuit, then the radio is still usable since mains power can be switched off and on at the supply socket; this mode of failure may thus be considered to be of low criticality. It is usually too tedious and time-consuming to take an FMECA down to component modes of failure, but this may be necessary in some special cases and particularly where hazardous modes of failure are possible. It may be necessary to consider multiple component failures in cases where the failure modes are complex or hazardous. The order in which failures occur can be significant: for example, failure of a control valve followed (a little later in time) by failure of an alarm circuit may not cause a dangerous condition since the failure will have been alarmed; the situation would be very different if the failures occurred in the opposite order.

Table 19.1 FMECA for a stereo cassette radio

Failure mode	*Effect*	*Criticality*
Power unit fails	Total system failure	1
Radio unit fails	Loss of radio	2
Cassette deck fails	Loss of ability to play or record on cassette	2
One audio channel fails	Loss of either left or right audio channel	3

Failure Mode, Effect Analysis (FMEA) is described in the following standard: IEC 812:1985 'Analysis techniques for system reliability – procedures for Failure Mode & Effects Analysis (FMEA), and is similar to FMECA except of course that no attempt is made to characterise the criticality of the failure mode. An FMECA, as well as being essential for the analysis of potentially dangerous failure modes, is also useful in maintenance planning because it can be used to identify dormant faults, see section 16.4.5. Once these are identified then regular preventive maintenance can be used to mitigate the problems which would be caused if the dormant faults were neglected indefinitely.

19.4 Use of Computer Programs

Computer programs are of great help in carrying out the more straightforward aspects of reliability analysis. However, the reader's attention is drawn to section 15.8 on the dangers of uninhibited use of programs.

A selective view of an ever-expanding range of computer software follows:

(1) *Programs which access component failure rate data and have facilities for the addition of failure rates.*

HardStress (HRD data) - ITEM Software Ltd
FRATE (HRD data) - British Telecom
BelStress (Bellcore data) - ITEM Software Ltd
Bellcore (Bellcore data) - Services Ltd
MilStress (MIL-HDBK-217 data) - ITEM Software Ltd
Predictor (MIL-HDBK-217 data) - Services Ltd
(2) *Programs for Monte Carlo Simulation/Availability Analysis*
Amir - Services Ltd
RAM 4 - Rex Thompson & Partners Ltd
Avail - ITEM Software Ltd
(3) *Programs for Spare Parts Provisioning*
SpareCost - ITEM Software Ltd
Spar - Services Ltd
(4) *Programs for FTA/FMECA*
FaultTree - ITEM Software Ltd
Treemaster - Services Ltd
Failure Mode, Effects and Criticality Analysis - Services Ltd
(5) *Programs on Maintenance/Life Cycle Costs*
Corida - Services Ltd
PPCM - Services Ltd
ILSA (Integrated Logistics Support Analysis) - ITEM Software Ltd

Addresses of the companies mentioned above are:
British Telecom, 81 Newgate Street, London EC1A 7AJ.
ITEM Software Ltd, Fearnside, Little Park Farm Road, Fareham, Hants. PO15 5SU.
Services Ltd, Quality and Reliability House, 82 Trent Boulevard, West Bridgford, Nottingham NG2 5BL.
Rex Thompson & Partners Ltd, 'Newnhams', West Street, Farnham, GU9 7EQ.

20 The Final Product

A good reliability engineer should have at least a background knowledge of statistics and probability theory. Starting from this background knowledge, this book attempts to provide the basic tools and methods for the practising engineer. There is an understandable tendency for the reliability engineer to become obsessed with numbers. After some considerable effort in identifying failure modes and calculating predicted MTTFs, he (and those he reports to) may feel that because his predictions are 'within specification' then the end product must be satisfactory. But the conscientious engineer will always try to remember that the goal is to produce a 'good' product. This goal (as seen by the buyer) is well defined by Mr G. E. Comer, Director of Reliability in the UK, Ministry of Defence (*Coderm Newsletter*, Vol. 7, No. 1, January 1990, published by the Coderm Secretariat, Archway South, Old Admiralty Building, Spring Gardens, London SW1A 2BE):

> 'Work on reliability is sometimes seen as being closely allied to mathematics and statistics. Although the numbers are important, they are only a means to an end. The object is to procure equipment which has been well designed and manufactured, which works and goes on working. Should it fail, it must be easily repaired.'

What is needed can be expressed in another way: 'The System Effectiveness must be high'. The concept of System Effectiveness was introduced at the beginning of this book and we return to it now, to recommend that the reliability engineer should always keep his eye on the ball, which in our case is labelled 'System Effectiveness'.

Finally, it should be clear from what has been written above that the design, manufacture, installation and maintenance of a reliable system calls for dedication from a very wide range of people: designers, production engineers, quality assurance and reliability engineers, installers, maintenance men and (most important of all) managers. Without the dedication of management to reliability (or, better, to System Effectiveness) much good work by engineers may be fruitless. The relative decline in engineering production in the UK in the years following the Second World War may be attributed to the lack of dedication of management to the concepts of Quality and Reliability. The best British managements have recognised the necessity of this commitment, and have started the up-hill task of strengthening British industry against outside competition. But much still remains to be done.

21 References

Abramowitz, M. and Stegun, I. A. (eds) (1964). *Handbook of Mathematical Functions with Formulas, Graphs and Mathematical Tables*, US Government Printing Office.

Aitken, J. and Brown, J. A. C. (1976). *The Lognormal Distribution*, Cambridge University Press.

Allen, A. O. (1978). *Probability, Statistics and Queuing Theory*, Academic Press.

Amerasekera, E. A. and Campbell, D. S. (1987). *Failure Mechanisms in Semiconductor Devices*, Wiley.

Bansal, V. K. *et al.* (1982). 'Minimal pathset and minimal cutsets using search technique', *Microelectronics and Reliability*, Vol. 22, No. 6, pp. 1067-1075.

Bazovsky, I. (1961). *Reliability Theory and Practice*, Prentice-Hall.

Bennetts, R. G. (1975). 'On the Analysis of fault trees', *IEEE Transactions on Reliability*, Vol. R-24, No. 3, August, pp. 175-185.

Cho, C-K. (1987). *Quality Programming: Developing and Testing Software with Statistical Quality Control*, Wiley.

Cox, D. R. (1962). *Renewal Theory*, Methuen.

Duane, J. T. (1964). 'Learning curve Approach to reliability modelling', *IEEE Transactions on Aerospace*, Vol. 2, No. 2, April.

Evans, R. A. (1969). 'Proper proof of a reliability theorem', *IEEE Transactions on Reliability*, November, pp. 205-206.

Evans, R. A. (1980). 'Independent or mutually exclusive' (editorial), *IEEE Transactions on Reliability*, Vol. R-29, No. 4, p. 289, October.

Evans, R. A. (1981). 'AQL is not PAQ!' (editorial), *IEEE Transactions on Reliability*, Vol. R-30, No. 3, August, p. 209.

Grosh, D. L. and Lyon, R. L. (1975). 'Stabilization of wearout-replacement rate', *IEEE Transactions on Reliability*, Vol. R-24, No. 4, October, pp. 268-270.

Hurley, R. B. (1963). 'Probability Maps', *IEEE Transactions on Reliability*, September, pp. 39-44.

Hill, F. J. and Peterson, G. R. (1981). *Introduction to Switching Theory and Logical Design*, 3rd edn, Wiley.

Jensen, F. and Peterson, N. E. (1982). *Burn-in: An Engineering Approach to the Design and Analysis of Burn-in Procedures*, Wiley.

Jeschke, C. E. *et al.* (1982). 'No. 10A remote switching system: physical design', *Bell System Technical Journal*, Vol. 61, No. 4, April, pp. 525-564.
Johnson, L. G. (1964). *Theory and Technique of Variation Research*, Elsevier, New York.
Johnson, L. R. (1969). 'Confidence interval estimation of the reliability of multi-component systems using component test data', Ph.D. Thesis in Mathematics, University of Delaware. (Obtainable through University Microfilms Inc., Ann Arbor, Michigan, USA.)
Kendall, M. G. and Stuart, A. (1973). *The Advanced Theory of Statistics*, Vol. 2, 3rd edn, Griffin.
Kitchenham, B. A. (1987). 'Management Metrics', *Proceedings of CSR Workshop on Reliability Achievement and Assessment 1987*, Blackwell Scientific.
Larsen, H. R. (1965). 'Nomographic binomial distribution', *Western Electric Engineer*, Vol. 9, April, pp. 20-29.
Larsen, H. R. (1966). 'A nomograph of the cumulative binomial distribution', *Industrial Quality Control*, December, pp. 270-278.
Leemis, L. M. (1988). 'Probabilistic properties of the exponential distribution', *Micro-electronics and Reliability*, Vol. 28, No. 2, pp. 257-262.
Lewis, E. E. (1987). *Introduction to Reliability Engineering*, Wiley.
Littlewood, B. and Verrall, J. L. 'A Bayesian reliability growth model for computer software', *Journal of the Royal Statistical Society - Series C*, Vol. 22, No. 3, pp. 332-346.
Mann, N. R., Shaeffer, R. E. and Singpurwalla, N. D. (1974). *Methods of Statistical Analysis of Reliability and Life Data*, Wiley.
Mead, P. H. (1977). 'Duane growth model: estimation of final MTBF with confidence limits using a hand calculator', *Proceedings of 1977 Annual Reliability and Maintainability Symposium*, pp. 269-274.
Musa, J. D., Iannino, A. and Okumoto, K. (1990). *Software Reliability: Measurement, Prediction, Application*, McGraw-Hill.
Musa, J. D. and Okumoto, K. (1984). 'Logarithmic Poisson execution time model for software reliability measurement', *Proceedings Seventh International Conference on Software Engineering, Orlando*, pp. 230-238.
Myers, G. J. (1976). *Software Reliability Principles and Practices*, Wiley.
O'Connor, P. D. T. (1985). *Practical Reliability Engineering*, 2nd edn, Wiley.
Pullum, G. G. and Grayson, H. (1979). 'The derivation and use of spare parts provisioning tables (taking into account the failure rate of units in storage)', *Proceedings of 2nd National Reliability Conference, Birmingham, U.K., March 1979*, Vol. 1, pp. 2D/1/1 to 2D/1/11.
Randell, B. (1975). 'System structure for software fault tolerance', *IEEE Transactions on Software Engineering*, Vol. SE-1, No. 2, June, pp. 220-232.
Roberts, N. C. (1988). 'An introduction to software reliability assessment', *Journal of the Safety and Reliability Society*, Vol. 8, No. 1, Spring, pp. 5-10.
Short, R. D. (1989). 'Sting ray: a sound-seeking missile', *IEE Review*, Vol. 35, No. 11, December, pp. 419-423.

Smith, D. J. (1985). *Reliability and Maintainability in Perspective*, 2nd edn, Macmillan.

Smith, D. J. (1986). 'Micro-electronic components failure rate data comparison', *9th Advances in Reliability Technology Symposium, Bradford (U.K.), April*, pp. C3/R/1 to C3/R/7.

Spencer, J. L. (1986). 'The highs and lows of reliability predictions', *Proceedings of Annual Reliability and Maintainability Symposium (IEEE)*, pp. 156–162.

Wald, A. (1947). *Sequential Analysis*, Wiley, New York.

Weik, M. K. (1983). *Communications Standard Dictionary*, Van Nostrand.

22 Bibliography

22.1 Reliability

1. Barlow, R. E. and Proschan, F., *Statistical Theory of Reliability and Life Testing - Probability Models*, Holt, Rinehart and Winston, 1975.
2. Carter, A. D. S., *Mechanical Reliability*, 2nd edn, Macmillan, 1986.
3. Green, A. E. and Bourne, A. J., *Reliability Technology*, Wiley, 1972.
4. Jardine, A. K. S., *Maintenance, Replacement and Reliability*, Pitman, 1973.
5. Kivenson, G., *Durability and Reliability in Engineering Design*, Pitman, 1972.
6. Lloyd, D. K. and Lipow, M., *Reliability - Management, Methods and Mathematics*, 2nd edn, American Society for Quality Control, Milwaukee, Wisconsin, 1984.
7. Mann, N. R. *et al.*, *Methods of Statistical Analysis of Reliability and Life Data*, Wiley, 1974.
8. O'Connor, P. D. T., *Practical Reliability Engineering*, 2nd edn, Wiley, 1985.
9. Shooman, M. L., *Probabilistic Reliability - an Engineering Approach*, McGraw-Hill, 1968.
10. Smith, D. J., *Reliability and Maintainability in Perspective*, 2nd edn, Macmillan, 1985.
11. Smith, D. J. and Babb, A. H., *Maintainability Engineering*, Pitman, 1973.

22.2 Probability and Statistics

1. Alder, H. L. and Roessler, E. B., *Introduction to Probability and Statistics*, 3rd edn, 1964; 5th edn, 1972, Freeman.

An easily readable book - very good on fundamentals. Not too mathematical.

2. Allen, A. O., *Probability, Statistics and Queueing Theory*, Academic Press, 1978.

Fairly mathematical: highly recommended.

3. Hoel, P. G., *Elementary Statistics*, 3rd edn, Wiley, 1971.
Hoel has also written a very good book called *Mathematical Statistics*.

4. Moroney, M. J., *Facts from Figures*, 3rd edn, Penguin, 1956.
A justly well-known book.

5. Weaver, W., *Lady Luck*, Penguin, 1977.
A very good book: complex material is treated in an easy-to-read manner.

6. Wine, R. L., *Statistics for Scientists and Engineers*, Prentice-Hall, 1964.
Rather old-fashioned but very good nevertheless.

23 Answers to Exercises

1.1. (a) For many people, the following factors may be considered to be the most important in buying a new car:

Life Cycle Cost
Technical Performance
Reliability
Comfort (Head-room, suspension, luggage space etc.)

But it must be remembered that the requirements will vary very widely with the individual. A very tall man might put Head-room above Life Cycle Cost if he has difficulty in finding a suitable car. A rich man might put a particular factor like 4-wheel drive at the top of his list, since he is unlikely to be worried about Life Cycle Cost.

(b) For a colour television set, likely factors in order of importance are:

Life Cycle Cost
Screen Size
Technical Performance
Reliability

1.2. In the absence of more detailed information, it is reasonable to consider the product

Technical Performance × Reliability

The three systems then have the following scores:

A - 0.64; B - 0.54; C - 0.465

It thus appears that system A is the most promising. In a real (rather than theoretical) situation, it would be as well to consider the capability of improvement in the Technical Performance and Reliability which may be achievable in each system before a final decision is made on which system should be developed.

3.1. 0.1 per cent/1000 hours $= 1 \times 10^{-3}/(1000 \text{ hours})$

$$= 1 \times 10^{-6}/\text{hour}$$
$$= 1000 \times 10^{-9}/\text{hour}$$
$$= 1000 \text{ fits}$$

1000 fits $= 1000 \times 10^{-9}/\text{hour}$

$$= 1000 \times 10^{-7}/(100 \text{ hours})$$
$$= 100 \times 1000 \times 10^{-7} \text{ per cent}/(100 \text{ hours})$$
$$= 0.01 \text{ per cent}/(100 \text{ hours})$$

3.2.

Time (hours)	*Expected number surviving* $n(t-5) \times 0.95\ (t > 5)$	$10\,000 \exp(-0.01t)$
0	10 000	10 000
5	9 500	9 512
10	9 025	9 048
15	8 574	8 607
20	8 145	8 187
25	7 738	7 788
30	7 351	7 408
35	6 983	7 047
40	6 634	6 703
45	6 302	6 376
50	5 987	6 065

Time (hours)	*Expected number surviving* $n(t-1) \times 0.99\ (t > 1)$	$10\,000 \exp(-0.01t)$
0	10 000	10 000
1	9 900	9 900
2	9 801	9 802
3	9 703	9 704
4	9 606	9 608
5	9 510	9 512

The discrepancy between the second stepped plot and the continuous curve $10\,000 \exp(-0.01t)$ is less than that of the first because the steps are smaller. As the steps become smaller and smaller, the stepped plot becomes closer and closer to the continuous curve. See Figure 23.1 on page 211. For the 1 hour steps, the two plots are indistinguishable.

3.3. Equating $R(t)$ with the proportion of non-decayed material:

$$0.5 = \exp(-\lambda t_1)$$

where λ has the dimensions of 1/hours. Inverting this expression:

$$2 = \exp(-\lambda t_1)$$

so that

$$\ln 2 = \lambda t_1$$

$$\lambda = \ln 2/t_1 \quad 1/\text{hours}$$

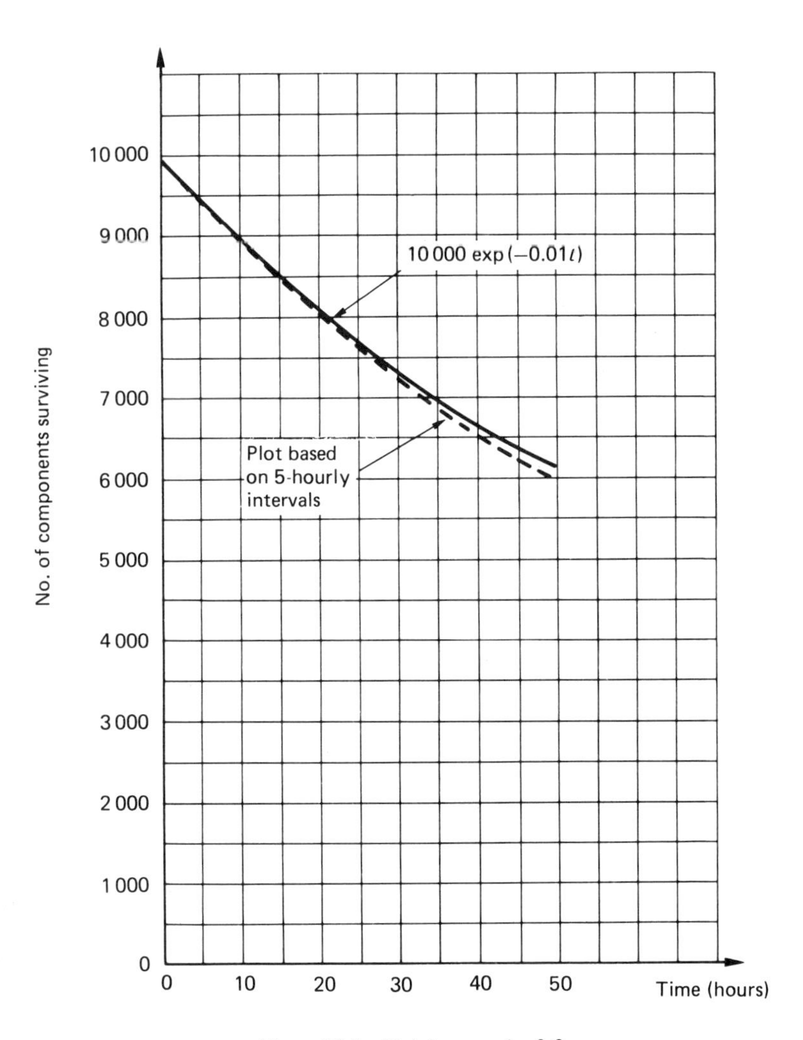

Figure 23.1 Plot for exercise **3.2**

3.4. Since the total area under the failure density curve must be unity, it follows that the height of the ordinate in Figure 3.15 must be $1/t_E$ (see Figure 23.2). Now $dR/dt = -f(t) = -1/t_E$ (a constant), and so $R(t)$ must have the shape shown in Figure 23.2 since it is also known that $R(0) = 1$. Since

$\lambda(t) = f(t)/R(t)$, it follows that $\lambda(0) = 1/t_E$ and that it must be a steadily increasing function as shown in Figure 23.2. Note that when $t = t_E/2$, $\lambda(t) = 2/t_E$; and when $t = 3t_E/4$, $\lambda(t) = 4/t_E$. The MTTF is the area under the $R(t)$ curve, which is $t_E/2$.

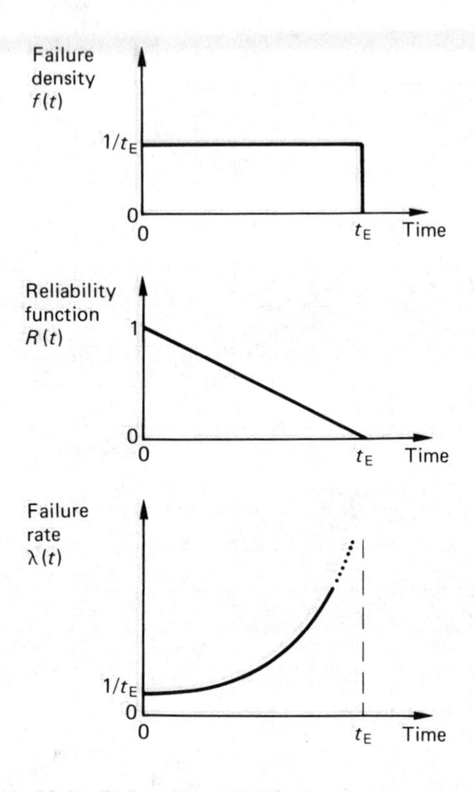

Figure 23.2 $R(t)$, $f(t)$ and $\lambda(t)$ for a certain component

4.1. Total component hours

$= 400 + 800 + 1200 + 1600 + 1900 + 39 \times 1900$
$= 80\,000$ hours

Since $k = 5$

$\hat{\theta} = 80\,000/5 = 16\,000$ hours

$\theta_{low} = 2T/_{0.10}\chi^2_{2k} = 160\,000/16.0 = 10\,000$ hours

4.2. Total component hours

$= 230 + 400 + 940 + 1250 + 6 \times 1300$
$= 10\,720$ hours

Since $k = 4$

$\hat{\theta} = 10\,720/4 = 2680$ hours

The test is time-truncated so

$$\theta_{low} = 2k\hat{\theta}/[_{0.05}\chi^2_{2(k+1)}] = 21\,440/18.3 = 1170 \text{ hours}$$

The upper confidence limit is given by

$$\theta_{upp} = 2k\hat{\theta}/_{0.95}\chi^2_{2k} = 21\,440/2.73 = 7850 \text{ hours}$$

4.3. For system (A): failure rate $= 1/\text{MTTF} = 1.0 \times 10^{-4}/\text{hour}$

For system (B): $\exp(-\lambda 1000) = 0.99$

So $\exp(\lambda 1000) = 1/0.99$

and $\lambda = [\ln(1/0.99)]/1000 = 1.0 \times 10^{-5}/\text{hour}$

You should choose system (B), since it has the lower failure rate.

4.4. Since there were no failures, the test must be treated as a time-truncated test. The lower 60 per cent single-sided confidence limit on the MTTF (given that the failure rate is constant) is given by

$$\theta_{low} = 2T/_{0.40}\chi^2_{2(k+1)}$$

so that in this case

$$\theta_{low} = 200\,000/_{0.40}\chi^2_2 = 200\,000/1.83 = 109\,000 \text{ hours}$$

$\hat{\theta}$ does not exist since T/k is undefined when $k = 0$. Similarly, θ_{upp} requires the calculation of $_{1-\alpha}\chi^2_{2k}$ which does not exist when $k = 0$.

4.5. Ten random numbers in the range 0 to 99 were taken from a set of tables. They were:

19, 38, 81, 50, 95, 52, 74, 33, 01, 32

When ordered from smallest to largest they become:

01, 19, 32, 33, 38, 50, 52, 74, 81, 95

Using the relationship $t_i = -\ln(1 - n_i)/\lambda$, the random times are obtained in the table overleaf (using $\lambda = 1 \times 10^{-4}/\text{hour}$).

On two-cycle logarithmic Weibull Probability Paper it is only possible to plot the times from 2100 hours upwards.

A reasonably good straight line is obtained with β approximately equal to unity.

Plots will of course vary, depending upon the random numbers which the reader selects. A lot of variability is possible and the reader should not be surprised if his plot shows more scatter than the one obtained here. It would not be surprising to obtain a value of β between 0.85 and 1.15.

Median rank (%)	n_i	$1 - n_i$	t_i *(hours)*
6.7	0.01	0.99	101
16.3	0.19	0.81	2 100
26.0	0.32	0.68	3 860
35.6	0.33	0.67	4 000
45.2	0.38	0.62	4 780
54.8	0.50	0.50	6 930
64.4	0.52	0.48	7 340
74.0	0.74	0.26	13 500
83.7	0.81	0.19	16 600
93.3	0.95	0.05	30 000

Owing to the assumed uniformity of the distribution of random numbers, the values n_i could have been used directly in the above table instead of $(1 - n_i)$.

4.6. It is shown in Appendix 1 that

$$f(\chi^2 : n) = \frac{1}{2^{n/2}\,\Gamma\left(\frac{n}{2}\right)} (\chi^2)^{\frac{n}{2}-1} \exp -\left(\frac{\chi^2}{2}\right)$$

When $n = 10$ this becomes

$$\frac{1}{2^5 \times 24} (\chi^2)^4 \exp[-\chi^2/2]$$

Values of χ^2 and corresponding values of $f(\chi^2 : 10)$ are:

χ^2	0	1	4	6	8	10	14	16	20
$f(\chi^2 : 10)$	0	7.9×10^{-4}	0.045	0.084	0.098	0.088	0.046	0.029	9.46×10^{-3}

Abscissas cutting off areas to the left and right of 0.20 are found from Appendix 1 to be 6.18 and 13.4 respectively. The plotted data should be identical with Figure 4.10 in the text.

4.7. The single-sided lower confidence limit on θ is given by

$$\theta = 2T/{}_{\alpha}\chi^2_{2(k+1)}$$

with a confidence level of $1 - \alpha$.

Therefore: 90 per cent lower single-sided limit is

$$2T/_{0.10}\chi^2_{12} = 200\,000/18.5 = 10\,800 \text{ hours}$$

80 per cent lower single-sided limit is

$$2T/_{0.20}\chi^2_{12} = 200\,000/15.8 = 12\,700 \text{ hours}$$

60 per cent lower single-sided limit is

$$2T/_{0.40}\chi^2_{12} = 200\,000/12.6 = 15\,900 \text{ hours}$$

Note: The above example shows how the confidence interval shrinks as the confidence level decreases. The point estimate of θ is 20 000 hours and is included in all of the above intervals. At 30 per cent confidence level (a level which would hardly ever be of interest), the confidence interval does not include the point estimate!

6.1. (a) System failure rate $= 100 \times 600 \times 10^9$
$= 6 \times 10^{-5}$ /hour

For a mission time of 1000 hours

system reliability = exp(–0.06) = 0.94

(b) We need a value of λ such that

$\exp(-100 \times \lambda \times 1000) = 0.98$

Solution of this equation gives

$\lambda = [\ln(1/0.98)]/100\,000 = 202 \times 10^{-9}$ /hour

(c) We need a value of t such that

$\exp(-100 \times \lambda \times t) = 0.98$

with $\lambda = 600 \times 10^{-9}$ /hour

Solution of this equation gives

$t = [\ln(1/0.98)]/[100 \times 600 \times 10^{-9}] = 337$ hours

6.2. Constant failure rates in units of 10^{-9} /hour:

Component	*Number*	*Total failure rate at temperature of* 25°C	85°C
Resistor	7	1.4	14
Capacitor	2	4	32
Transistor	2	40	88
Solder Joint	24	1.7	1.7
TOTAL		47.1	135.7
MTTF (hours)		21.2×10^6	7.37×10^6
Probability of survival for 8000 hours		0.9996	0.9989

6.3. When λ_b = 10 000 fits:

Component type	*Number used*	*Failure rate (fits)*	*(No. used) × (fail. rate)*	*Percentage contribution to total*
a	3	10	30	0
b	4	10 000	40 000	67
c	10	1 000	10 000	17
d	7	850	5 950	10
e	2	2 000	4 000	7
TOTAL			59 980	

MTTF = 16 700 hours

When λ_b = 5000 fits:

Component type	*Number used*	*Failure rate (fits)*	*(No. used) × (fail. rate)*	*Percentage contribution to total*
a	3	10	30	0
b	4	5 000	20 000	50
c	10	1 000	10 000	25
d	7	850	5 950	15
e	2	2 000	4 000	10
TOTAL			39 980	

MTTF = 25 000 hours

Percentage contribution of component type b to total failure rate falls from 67 per cent to 50 per cent.

8.1. The probability that a single fan will survive a mission time of 10 000 hours is 0.9, and thus the probability of failure is 0.1.
The possible states of this system and their probabilities are the five terms in the binomial expansion of $(0.9 + 0.1)^4$, since the probabilities of failure are independent.
Thus:

probability that all 4 fans survive mission time $= 0.9^4 = 0.6561$
probability that 3 fans survive mission time $= 4 \times 0.9^3 \times 0.1 = 0.2916$
probability that 2 fans survive mission time $= 6 \times 0.9^2 \times 0.1^2 = 0.0486$
probability that 1 fan survives mission time $= 4 \times 0.9 \times 0.1^3 = 0.0036$
probability that 0 fans survive mission time $= 0.1^4 = 0.0001$

Probability that the mission will be completed successfully

$$= 0.6561 + 0.2916 + 0.0486 = 0.9963$$

8.2. For system (A): $R_A(t) = 2\exp(-\lambda t) - \exp(-2\lambda t)$
where λ is the failure rate of one of the (identical) parallel units.
Thus, with t = 1000 hours, and $\lambda = 1 \times 10^{-4}$/hour:

$$R_A(1000) = 0.991$$

Mean time to failure = $3/2\lambda$ = 15 000 hours.
For system (B): $R_B = \exp(-\lambda t)$
Thus, with t = 1000 hours and $\lambda = 5 \times 10^{-5}$/hour:

$$R_B(1000) = 0.951$$

Mean time to failure = $1/\lambda$ = 20 000 hours.
Thus system (B) has the higher mean time to failure. System (A) has a higher probability of surviving a 1000 hour mission.

9.1. Let S be the event that the system is 'up'. Using the Theorem of Total Probabilities, select d as the pivotal unit. Then

$$R_s = \Pr(S \mid d)R_d + \Pr(S \mid \bar{d})F_d$$

By inspection

$$\Pr(S \mid d) = R_a + R_b - R_aR_b$$

which is the reliability of units a and b in parallel, and

$$\Pr(S \mid \bar{d}) = R_aR_c + R_bR_e - R_aR_bR_cR_e$$

which is the reliability of units a and c (in series) in parallel with b and e (in series). Thus

$$R_s = (R_a + R_b - R_aR_b)R_d + (R_aR_c + R_bR_e - R_aR_bR_cR_e)(1 - R_d)$$

9.2. With unit c as the pivotal unit:

$$R_s = \Pr(S \mid c)R_c + \Pr(S \mid \bar{c})F_c$$

If unit c is 'up', then $\Pr(S \mid c)$ becomes the reliability of units a and b in parallel:

$$\Pr(S \mid c) = R_a + R_b - R_aR_b$$

If unit c is down, then $\Pr(S \mid \bar{c})$ becomes the reliability of units b and d in series:

$$\Pr(S \mid \bar{c}) = R_bR_d$$

Hence

$$R_s = (R_a + R_b - R_aR_b)R_c + R_bR_d\,(1 - R_c)$$
$$= R_aR_c\,(1 - R_b) + R_b(R_c + R_d - R_cR_d)$$

10.1. $L\lambda \exp(-\lambda t) = \lambda/(s + \lambda)$

and similarly

$$L\mu \exp(-\mu t) = \mu/(s + \mu)$$

Hence the product of the Laplace transforms is

$$\lambda\mu/[(s + \lambda)(s + \mu)]$$

This has an inverse of $\mu\lambda[\exp(-\lambda t) - \exp(-\mu t)]/(\mu - \lambda)$ which is the density of the cycle time. Since the mean of $\lambda \exp(-\lambda t)$ is $1/\lambda$, with a corresponding expression in μ, it is easy to see that the mean cycle time is $[\mu/\lambda - \lambda/\mu]/(\mu - \lambda) = 1/\lambda + 1/\mu$.

10.2. $\int_0^{t_R} R(t)\mathrm{d}t = \int_0^{t_R} (1 - t/t_E)\mathrm{d}t = [t - t^2/2t_E]_0^{t_R} = t_R(1 - t_R/2t_E)$

Also $R(t_R) = 1 - t_R/t_E$
Hence Positional Mean Up Time $= t_R(1 - t_R/2t_E)/(1 - t_R/t_E)$
Putting $t_R = 10\,000$ hours and $t_E = 20\,000$ hours gives

$$\begin{aligned} \text{PMUT} &= 10\,000\,(1 - 1/4)/(1 - 1/2) \\ &= 15\,000 \text{ hours} \end{aligned}$$

MTTF = 10 000 hours (see Exercise **3.4**)

11.1.

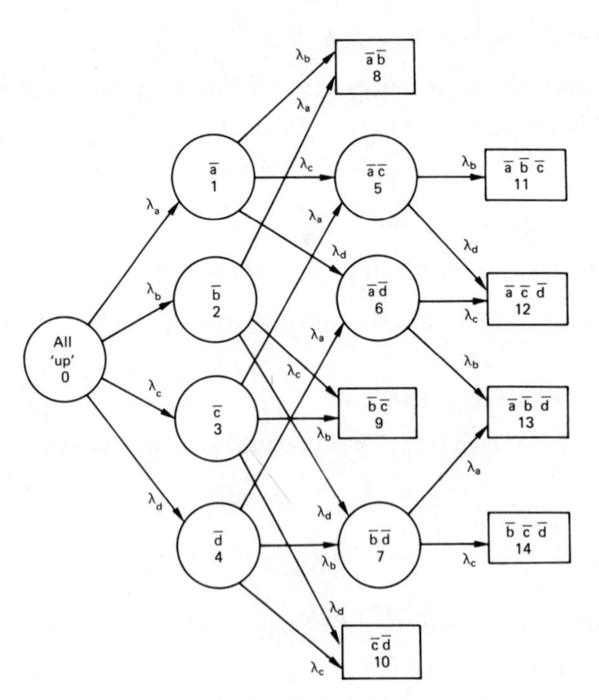

Figure 23.3 Markov diagram for exercise **11.1**

11.2. Using the results of section 11.9:

$$\begin{aligned}\text{MUT} &= 1/(\lambda_a + \lambda_b) = 1/(150\,000 \text{ fits}) \\ &= 10/150\,000 \text{ hours} \\ &= 6700 \text{ hours}\end{aligned}$$

$$\begin{aligned}\text{MDT} &= [\lambda_a/(\lambda_a + \lambda_b)]\,(1/\mu_a) + [\lambda_b/(\lambda_a + \lambda_b)]\,(1/\mu_b) \\ &= (1/1.5)10 + (0.5/1.5)50 = 23.3 \text{ hours}\end{aligned}$$

11.3. The modified Markov diagram is shown in Figure 23.4.

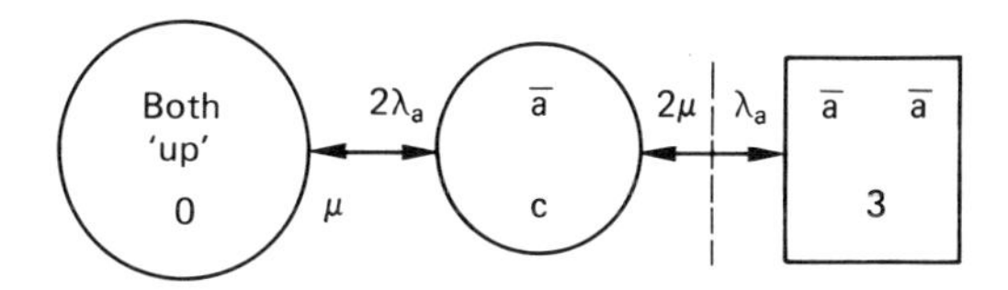

Figure 23.4 Markov diagram for two identical units in parallel and as many repairmen as faults

The equations to be solved are:

$$\begin{aligned}0 &= -2\lambda_a P_0(\infty) + \mu P_c(\infty) \\ 0 &= 2\lambda_a P_0(\infty) - (\lambda_a + \mu)P_c(\infty) + 2\mu P_3(\infty) \\ 0 &= \lambda_a P_c(\infty) - 2\mu P_3(\infty) \\ 1 &= P_0(\infty) + P_c(\infty) + P_3(\infty)\end{aligned}$$

which gives:

$$P_0(\infty) = 1/D_1 \qquad P_c(\infty) = \frac{2\lambda_a}{\mu}\Big/D_1 \qquad P_3(\infty) = (\lambda_a^2/\mu^2)/D_1$$

where $D_1 = 1 + 2\lambda_a/\mu + \lambda_a^2/\mu^2$.
The steady-state unavailability is

$$P_3(\infty) = (\lambda_a^2/\mu^2)/D_1$$

The corresponding unavailability for the system of Figure 11.8 is

$$(2\lambda_a^2/\mu)/D$$

where $D = 1 + 2\lambda_a/\mu + 2\lambda_a^2/\mu^2$.

11.4. The Markov diagram is as shown in Figure 23.5.

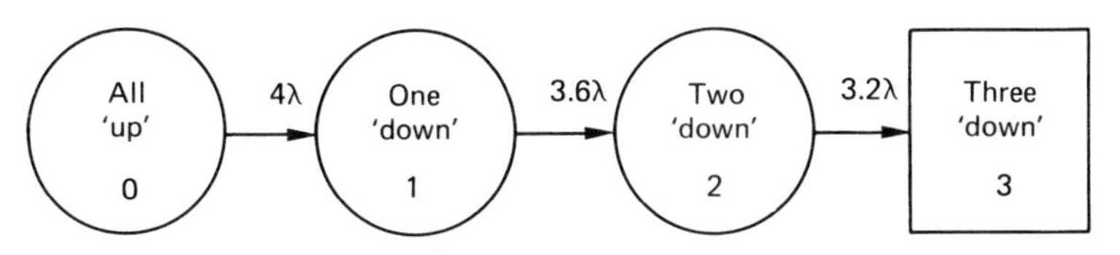

Figure 23.5 Markov diagram for Exercise **11.4**

Differential equations to be solved are

$$
\begin{aligned}
dP_0/dt &= -4\lambda P_0(t) \\
dP_1/dt &= 4\lambda P_0(t) - 3.6\lambda P_1(t) \\
dP_2/dt &= 3.6\lambda P_1(t) - 3.2\lambda P_2(t) \\
dP_3/dt &= 3.2\lambda P_2(t)
\end{aligned}
$$

Laplace transforms of these equations are

$$
\begin{aligned}
s\bar{P}_0(s) - 1 &= -4\lambda\bar{P}_0(s) \\
s\bar{P}_1(s) &= 4\lambda\bar{P}_0(s) - 3.6\lambda\bar{P}_1(s) \\
s\bar{P}_2(s) &= 3.6\lambda\bar{P}_1(s) - 3.2\lambda\bar{P}_2(s) \\
s\bar{P}_3(s) &= 3.2\lambda\bar{P}_2(s)
\end{aligned}
$$

which give

$$\bar{P}_0(s) = 1/(s + 4\lambda)$$

$$\bar{P}_1(s) = 4\lambda\bar{P}_0(s)/(s + 3.6\lambda) = 4\lambda/[(s + 4\lambda)(s + 3.6\lambda)]$$

$$\bar{P}_2(s) = 3.6\lambda\bar{P}_1(s)/(s + 3.2\lambda) = 14.4\lambda^2/[(s + 4\lambda)(s + 3.6\lambda)(s + 3.2\lambda)]$$

$$\bar{P}_3(s) = 3.2\lambda\bar{P}_2(s)/s = 46.08\lambda^3/[s(s + 4\lambda)(s + 3.6\lambda)(s + 3.2\lambda)]$$

Inversion of $\bar{P}_3(s)$ gives

$$P_3(t) = 1 - 36\exp(-4\lambda t) + 80\exp(-3.6\lambda t) - 45\exp(-3.2\lambda t)$$

The aircraft reliability is thus

$$R(t) = 1 - P_3(t) = 36\exp(-4\lambda t) - 80\exp(-3.6\lambda t) + 45\exp(-3.2\lambda t)$$

The probability of completing a seven-hour mission is

$$R(t) = 36\exp(-28\lambda) - 80\exp(-25.2\lambda) + 45\exp(-22.4\lambda)$$

where λ must have the dimensions of hour^{-1}.

12.1. Using $m = 10$, $\lambda = 1/1000/\text{hour}$, $T = 200$ hours gives

$$m\lambda T = 2$$

Putting $\Pr(i)$ to stand for the probability of i failures gives

$$\Pr(0) = \exp(-m\lambda T) = 0.1353$$

$$\Pr(1) = m\lambda T\exp(-m\lambda T)/1! = 0.2707$$

Hence the probability that stock will not be exhausted is

$$\Pr(0) + \Pr(1) = 0.4060$$

Probability of stock-out = 1 – 0.4060 = 0.5940.

12.2. The Markov diagram is shown below.

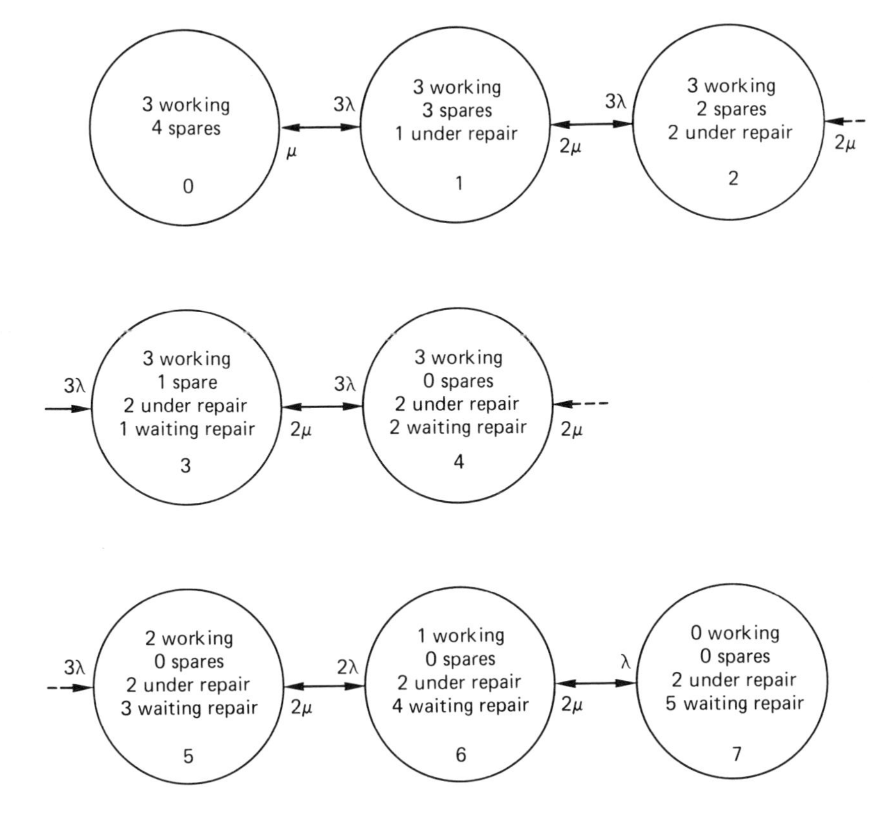

Figure 23.6 Markov diagram for Exercise **12.2**

15.1. The Markov diagram is shown in Figure 23.7 where

$$\lambda_u = \lambda_a + \lambda_{ch} + \lambda_d$$

Differential equations obtained from this diagram are

$$\begin{aligned}
dP_0/dt &= -(\lambda_u + 2\lambda_b)P_0 + \mu_h P_1 + \mu_h P_2 \\
dP_1/dt &= 2\lambda_b P_0 - (\mu_h + \lambda_b)P_1 + 2\mu_h P_3 \\
dP_2/dt &= \lambda_u P_0 - \mu_h P_2 \\
dP_3/dt &= \lambda_b P_1 - 2\mu_h P_3
\end{aligned}$$

Setting the left-hand sides of these equations to zero and solving for the P terms yields

$$P_0(\infty) = 1/D \qquad P_1(\infty) = (2\lambda_b/\mu_h)(1/D)$$
$$P_2(\infty) = (\lambda_u/\mu_h)(1/D) \qquad P_3(\infty) = (\lambda_b/\mu_h)^2(1/D)$$

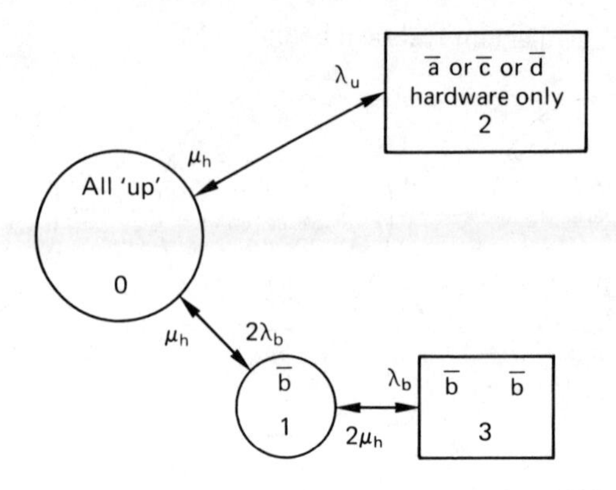

Figure 23.7 Markov diagram for Exercise **15.1**

where

$$D = 1 + (2\lambda_b + \lambda_u)/\mu_h + (\lambda_b/\mu_h)^2$$

Then

$$MCT_h = 1/(\lambda_u P_0(\infty) + \lambda_b P_1(\infty)) = D/(\lambda_u + 2\lambda_b^2/\mu_h)$$

$$U_{ssh} = P_2(\infty) + P_3(\infty) = [\lambda_u/\mu_h + (\lambda_b/\mu_h)^2]/D$$

$$MDT_h = U_{ssh}\, MCT_h = [\lambda_u/\mu_h + (\lambda_b/\mu_h)^2]/(\lambda_u + 2\lambda_b^2/\mu_h)$$

$$= \frac{\lambda_u}{\lambda_u + 2\lambda_b^2/\mu_h} \cdot \frac{1}{\mu_h} + \frac{2\lambda_b^2/\mu_h}{\lambda_u + 2\lambda_b^2/\mu_h} \cdot \frac{1}{2\mu_h}$$

Note: It would be incorrect to collapse states 2 and 3 as reference to the equations for dP_2/dt and dP_3/dt shows.

15.2. The Markov diagram of 23.7 becomes Figure 23.8 when the possibility of software failure is added.

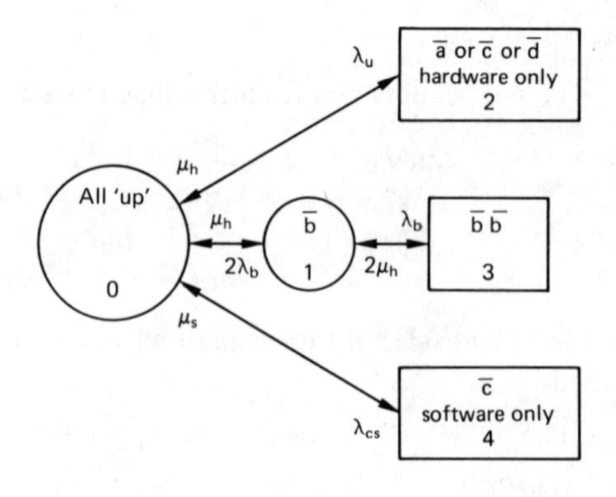

Figure 23.8 Markov diagram for Exercise **15.2**

Differential equations for this diagram are

$$\begin{aligned}
dP_0/dt &= -(\lambda_u + 2\lambda_b + \lambda_{cs})P_0 + \mu_h P_1 + \mu_h P_2 + \mu_s P_4 \\
dP_1/dt &= 2\lambda_b P_0 - (\mu_h + \lambda_b)P_1 + 2\mu_h P_4 \\
dP_2/dt &= \lambda_u P_0 - \mu_h P_2 \\
dP_3/dt &= \lambda_b P_1 - 2\mu_h P_4 \\
dP_4/dt &= \lambda_{cs} P_0 - \mu_s P_4
\end{aligned}$$

Setting the left-hand sides of these equations to zero and solving for the P terms yields

$$P_0(\infty) = 1/D \qquad P_1(\infty) = (2\lambda_b/\mu_h)/D$$
$$P_2(\infty) = (\lambda_u/\mu_h)/D \qquad P_3(\infty) = (\lambda_b/\mu_h)^2/D$$
$$P_4(\infty) = (\lambda_{cs}/\mu_s)/D$$

where

$$D = 1 + (2\lambda_b + \lambda_u)/\mu_h + (\lambda_b/\mu_h)^2 + \lambda_{cs}/\mu_s$$

Then

$$\begin{aligned}
\text{MCT} &= 1/[(\lambda_u + \lambda_{cs})P_0(\infty) + \lambda_b P_1(\infty)] \\
&= D/[\lambda_u + \lambda_{cs} + 2\lambda_b^2/\mu_h] \\
U_{ss} &= P_2(\infty) + P_3(\infty) + P_4(\infty) = [\lambda_u/\mu_h + (\lambda_b/\mu_h)^2 + \lambda_{cs}/\mu_s]/D
\end{aligned}$$

and

$$\begin{aligned}
\text{MDT} &= U_{ss}\,\text{MCT} \\
&= [\lambda_u/\mu_h + (\lambda_b/\mu_h)^2 + \lambda_{cs}/\mu_s]/[\lambda_u + \lambda_{cs} + 2\lambda_b^2/\mu_h]
\end{aligned}$$

Putting $\lambda_{hw} = \lambda_u + 2\lambda_b^2/\mu_h$, $\lambda_{sw} = \lambda_{cs}$ and re-arranging yields

$$\begin{aligned}
\text{MDT} &= \frac{(\lambda_u + \lambda_b^2/\mu_h)(1/\mu_h)}{\lambda_{hw} + \lambda_{sw}} + \frac{\lambda_{sw}(1/\mu_s)}{\lambda_{hw} + \lambda_{sw}} \\
&= \frac{\lambda_{hw}}{\lambda_{hw} + \lambda_{sw}} \cdot \frac{(\lambda_u + \lambda_b^2/\mu_h)(1/\mu_h)}{\lambda_{hw}} + \frac{\lambda_{sw}(1/\mu_s)}{\lambda_{hw} + \lambda_{sw}} \\
&= \frac{\lambda_{hw}}{\lambda_{hw} + \lambda_{sw}}\,\text{MDT}_h + \frac{\lambda_{sw}}{\lambda_{hw} + \lambda_{sw}}\,\text{MDT}_s
\end{aligned}$$

Calculation of MUT:

$$A_{ss} = P_0(\infty) + P_1(\infty) = (1 + 2\lambda_b/\mu_h)/D$$

$$\begin{aligned}
\text{MUT} = A_{ss}.\ \text{MCT} &= (1 + 2\lambda_b/\mu_h)/[\lambda_u + \lambda_{cs} + 2\lambda_b^2/\mu_h] \\
&= 1.01/(50\,000 \text{ fits}) \\
&= 20\,200 \text{ hours}
\end{aligned}$$

The approximate MUT calculated in the text is thus within 1 per cent of the accurate answer.

Appendix 1: χ^2 Distribution Tables for Even-numbered Degrees of Freedom

The tables give values of χ^2 such that the area cut off to the right of the abscissa ${}_{\gamma}\chi^2_n$ is γ $(0 \leqslant \gamma \leqslant 1)$, that is

$$\gamma = \int_{{}_{\gamma}\chi^2_n}^{\infty} f(\chi^2 : n)\,\mathrm{d}\chi^2$$

as illustrated in Figure A1.1. For reliability purposes, even-numbered degrees of freedom only are needed; for this reason, odd-numbered degrees of freedom are omitted from the tables.

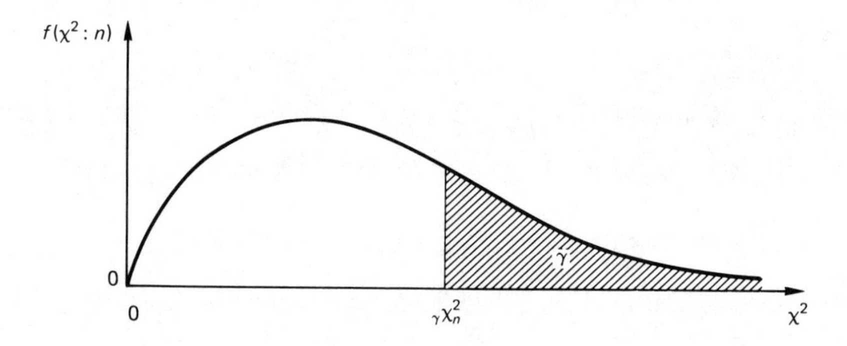

Figure A1.1 The χ^2 distribution (n degrees of freedom)

Note that

$$f(\chi^2 : n) = \frac{1}{2^{n/2}\Gamma(n/2)}\,(\chi^2)^{\frac{n}{2}-1}\,\exp(-\chi^2/2)$$

see Abramowitz & Stegun (1964).
Since the χ^2 distribution is a probability density function it follows that

$$\int_0^{\infty} f(\chi^2 : n)\mathrm{d}\chi^2 = 1$$

Percentage points of χ^2 distribution

n \ γ	0.9995	0.999	0.995	0.990	0.975	0.95	0.90	0.80	0.70	0.60
2	1.00E−3	2.00E−3	0.0100	0.0201	0.0506	0.103	0.211	0.466	0.713	1.02
4	0.0639	0.0908	0.207	0.297	0.484	0.711	1.06	1.65	2.19	2.75
6	0.299	0.381	0.676	0.872	1.24	1.64	2.20	3.07	3.83	4.57
8	0.710	0.857	1.34	1.65	2.18	2.73	3.49	4.59	5.53	6.42
10	1.26	1.48	2.16	2.56	3.25	3.94	4.87	6.18	7.27	8.30
12	1.93	2.21	3.07	3.57	4.40	5.23	6.30	7.81	9.03	10.2
14	2.70	3.04	4.07	4.66	5.63	6.57	7.79	9.47	10.8	12.1
16	3.54	3.94	5.14	5.81	6.91	7.96	9.31	11.2	12.6	14.0
18	4.44	4.90	6.26	7.01	8.23	9.39	10.9	12.9	14.4	15.9
20	5.40	5.92	7.43	8.26	9.59	10.9	12.4	14.6	16.3	17.8
22	6.40	6.98	8.64	9.54	11.0	12.3	14.0	16.3	18.1	19.7
24	7.45	8.08	9.89	10.9	12.4	13.8	15.7	18.1	19.9	21.7
26	8.45	9.22	11.2	12.2	13.8	15.4	17.3	19.8	21.8	23.6
28	9.66	10.4	12.5	13.6	15.3	16.9	18.9	21.6	23.6	25.5
30	10.8	11.6	13.8	15.0	16.8	18.5	20.6	23.4	25.5	27.4
32	12.0	12.8	15.1	16.4	18.3	20.1	22.3	25.1	27.4	29.4
34	13.2	14.1	16.5	17.8	19.8	21.7	24.0	26.9	29.2	31.3
36	14.4	15.3	17.9	19.2	21.3	23.3	25.6	28.7	31.1	33.3
38	15.6	16.6	19.3	20.7	22.9	24.9	27.3	30.5	33.0	35.2
40	16.9	17.9	20.7	22.2	24.4	26.5	29.1	32.3	34.9	37.1
42	18.2	19.2	22.1	23.7	26.0	28.1	30.8	34.2	36.8	39.1
44	19.5	20.6	23.6	25.1	27.6	29.8	32.5	36.0	38.6	41.0
46	20.8	21.9	25.0	26.7	29.1	31.4	34.2	37.8	40.5	43.0
48	22.1	23.3	26.5	28.2	30.8	33.1	35.9	39.6	42.4	44.9
50	23.5	24.7	28.0	29.7	32.4	34.8	37.7	41.4	44.3	46.9

Percentage points of χ^2 distribution

0.50	0.40	0.30	0.20	0.10	0.05	0.025	0.01	0.005	0.001	0.0005	γ / n
1.39	1.83	2.41	3.22	4.61	5.99	7.38	9.21	10.6	13.8	15.2	2
3.36	4.04	4.88	5.99	7.78	9.49	11.1	13.3	14.9	18.5	20.0	4
5.35	6.21	7.23	8.56	10.6	12.6	14.4	16.8	18.5	22.5	24.1	6
7.34	8.35	9.52	11.0	13.4	15.5	17.5	20.1	22.0	26.1	27.9	8
9.34	10.5	11.8	13.4	16.0	18.3	20.5	23.2	25.2	29.6	31.4	10
11.3	12.6	14.0	15.8	18.5	21.0	23.3	26.2	28.3	32.9	34.8	12
13.3	14.7	16.2	18.2	21.1	23.7	26.1	29.1	31.3	36.1	38.1	14
15.3	16.8	18.4	20.5	23.5	26.3	28.8	32.0	34.3	39.3	41.3	16
17.3	18.9	20.6	22.8	26.0	28.9	31.5	34.8	37.2	42.3	44.4	18
19.3	21.0	22.8	25.0	28.4	31.4	34.2	37.6	40.0	45.3	47.5	20
21.3	23.0	24.9	27.3	30.8	33.9	36.8	40.3	42.8	48.3	50.5	22
23.3	25.1	27.1	29.6	33.2	36.4	39.4	43.0	45.6	51.2	53.5	24
25.3	27.2	29.2	31.8	35.6	38.9	41.9	45.6	48.3	54.1	56.4	26
27.3	29.2	31.4	34.0	37.9	41.3	44.5	48.3	51.0	56.9	59.3	28
29.3	31.3	33.5	36.3	40.3	43.8	47.0	50.9	53.7	59.7	62.2	30
31.3	33.4	35.7	38.5	42.6	46.2	49.5	53.5	56.3	62.5	65.0	32
33.3	35.4	37.8	40.7	44.9	48.6	52.0	56.1	59.0	65.2	67.8	34
35.3	37.5	39.9	42.9	47.2	51.0	54.4	58.6	61.6	68.0	70.6	36
37.3	39.6	42.0	45.1	49.5	53.4	56.9	61.2	64.2	70.7	73.4	38
39.3	41.6	44.2	47.3	51.8	55.8	59.3	63.7	66.8	73.4	76.1	40
41.3	43.7	46.3	49.5	54.1	58.1	61.8	66.2	69.3	76.1	78.8	42
43.3	45.7	48.4	51.6	56.4	60.5	64.2	68.7	71.9	78.7	81.5	44
45.3	47.8	50.5	53.8	58.6	62.8	66.6	71.2	74.4	81.4	84.2	46
47.3	49.8	52.6	56.0	60.9	65.2	69.0	73.7	77.0	84.0	86.9	48
49.3	51.9	54.7	58.2	63.2	67.5	71.4	76.2	79.5	86.7	89.6	50

Percentage points of χ^2 distribution

n \ γ	0.9995	0.999	0.995	0.990	0.975	0.95	0.90	0.80	0.70	0.60
52	24.8	26.1	29.5	31.2	34.0	36.4	39.4	43.3	46.2	48.8
54	26.2	27.5	31.0	32.8	35.6	38.1	41.2	45.1	48.1	50.8
56	27.6	28.9	32.5	34.3	37.2	39.8	42.9	47.0	50.0	52.7
58	28.9	30.3	34.0	35.9	38.8	41.5	44.7	48.8	51.9	54.7
60	30.3	31.7	35.5	37.5	40.5	43.2	46.5	50.6	53.8	56.6
62	31.7	33.2	37.1	39.1	42.1	44.9	48.2	52.5	55.7	58.6
64	33.2	34.6	38.6	40.6	43.8	46.6	50.0	54.3	57.6	60.5
66	34.6	36.1	40.2	42.2	45.4	48.3	51.8	56.2	59.5	62.5
68	36.0	37.5	41.7	43.8	47.1	50.0	53.5	58.0	61.4	64.4
70	37.5	39.0	43.3	45.4	48.8	51.7	55.3	59.9	63.3	66.4
72	38.9	40.5	44.8	47.1	50.4	53.5	57.1	61.8	65.3	68.4
74	40.4	42.0	46.4	48.7	52.1	55.2	58.9	63.6	67.2	70.3
76	41.8	43.5	48.0	50.3	53.8	56.9	60.7	65.5	69.1	72.3
78	43.3	45.0	49.6	51.9	55.5	58.7	62.5	67.3	71.0	74.2
80	44.8	46.5	51.2	53.5	57.2	60.4	64.3	69.2	72.9	76.2
82	46.3	48.0	52.8	55.2	58.8	62.1	66.1	71.1	74.8	78.1
84	47.8	49.6	54.4	56.8	60.5	63.9	67.9	72.9	76.8	80.1
86	49.3	51.1	56.0	58.5	62.2	65.6	69.7	74.8	78.7	82.1
88	50.8	52.6	57.6	60.1	63.9	67.4	71.5	76.7	80.6	84.0
90	52.3	54.2	59.2	61.8	65.6	69.1	73.3	78.6	82.5	86.0
92	53.8	55.7	60.8	63.4	67.4	70.9	75.1	80.4	84.4	88.0
94	55.3	57.2	62.4	65.1	69.1	72.6	76.9	82.3	86.4	89.9
96	56.8	58.8	64.1	66.7	70.8	74.4	78.7	84.2	88.3	91.9
98	58.4	60.4	65.7	68.4	72.5	76.2	80.5	86.1	90.2	93.8
100	59.9	61.9	67.3	70.1	74.2	77.9	82.4	87.9	92.1	95.8

Percentage points of χ^2 distribution

0.50	0.40	0.30	0.20	0.10	0.05	0.025	0.01	0.005	0.001	0.0005	γ / n
51.3	53.9	56.8	60.3	65.4	69.8	73.8	78.6	82.0	89.3	92.2	52
53.3	56.0	58.9	62.5	67.7	72.2	76.2	81.1	84.5	91.9	94.8	54
55.3	58.0	61.0	64.7	69.9	74.5	78.6	83.5	87.0	94.5	97.5	56
57.3	60.1	63.1	66.8	72.2	76.8	80.9	86.0	89.5	97.0	100.1	58
59.3	62.1	65.2	69.0	74.4	79.1	83.3	88.4	92.0	99.6	102.7	60
61.3	64.2	67.3	71.1	76.6	81.4	85.7	90.8	94.4	102.2	105.3	62
63.3	66.2	69.4	73.3	78.9	83.7	88.0	9[illegible].2	96.9	104.7	107.9	64
65.3	68.3	71.5	75.4	81.1	86.0	90.3	95.6	99.3	107.3	110.5	66
67.3	70.3	73.6	77.6	83.3	88.3	92.7	98.0	101.8	109.8	113.0	68
69.3	72.4	75.7	79.7	85.5	90.5	95.0	100.4	104.2	112.3	115.6	70
71.3	74.4	77.8	81.9	87.7	92.8	97.4	102.8	106.6	114.8	118.1	72
73.3	76.4	79.9	84.0	90.0	95.1	99.7	105.2	109.1	117.3	120.7	74
75.3	78.5	82.0	86.1	92.2	97.4	102.0	107.6	111.5	119.9	123.2	76
77.3	80.5	84.0	88.3	94.4	99.6	104.3	110.0	113.9	122.3	125.7	78
79.3	82.6	86.1	90.4	96.6	101.9	106.6	112.3	116.3	124.8	128.3	80
81.3	84.6	88.2	92.5	98.8	104.1	108.9	114.7	118.7	127.3	130.8	82
83.3	86.6	90.3	94.7	101.0	106.4	111.2	117.1	121.1	129.8	133.3	84
85.3	88.7	92.4	96.8	103.2	108.6	113.5	119.4	123.5	132.3	135.8	86
87.3	90.7	94.4	98.9	105.4	110.9	115.8	121.8	125.9	134.7	138.3	88
89.3	92.8	96.5	101.1	107.6	113.1	118.1	124.1	128.3	137.2	140.8	90
91.3	94.8	98.6	103.2	109.8	115.4	120.4	126.5	130.7	139.7	143.3	92
93.3	96.8	100.7	105.3	111.9	117.6	122.7	128.8	133.1	142.1	145.8	94
95.3	98.9	102.8	107.4	114.1	119.9	125.0	131.1	135.4	144.6	148.2	96
97.3	100.9	104.8	109.5	116.3	122.1	127.3	133.5	137.8	147.0	150.7	98
99.3	102.9	106.9	111.7	118.5	124.3	129.6	135.8	140.2	149.4	153.2	100

Values of *n* greater than 100

When n is greater than 100, the following approximation may be used for χ^2:

$${}_{\gamma}\chi_n^2 \approx \tfrac{1}{2}\,[z_\gamma + \sqrt{(2n-1)}]^2$$

where z_γ is given in the following table:

γ	0.9995	0.999	0.995	0.990	0.975	0.95	0.90
z_γ	−3.2905	−3.0902	−2.5758	−2.3263	−1.9600	−1.6449	−1.2816

γ	0.80	0.70	0.60	0.50	0.40	0.30	0.20
z_γ	−0.8416	−0.5244	−0.2533	0	0.2533	0.5244	0.8416

γ	0.10	0.05	0.025	0.01	0.005	0.001	0.0005
z_γ	1.2816	1.6449	1.9600	2.3263	2.5758	3.0902	3.2905

Appendix 2: 5%, 95%, and Median (50%) Rank Tables

The general method for obtaining ranks is:

the δ per cent for the ith rank order out of a sample size of n is obtained by solving the equation

$$q^n + {}_nC_{n-1}q^{n-1}(1-q) + \ldots + {}_nC_i q^i (1-q)^{n-i} = \delta/100$$

The required rank (as a percentage plotting point) is then $100q$.
Mean ranks (as percentages) are

$$100i/(n+1)$$

5 per cent ranks

Rank order	*Sample size* 1	2	3	4	5	6	7	8	9	10
1	5.0	2.5	1.7	1.3	1.0	0.9	0.7	0.6	0.6	0.5
2		22.4	13.5	9.8	7.6	6.3	5.3	4.6	4.1	3.7
3			36.8	24.9	18.9	15.3	12.9	11.1	9.8	8.7
4				47.3	34.3	27.1	22.5	19.3	16.9	15.0
5					54.9	41.8	34.1	28.9	25.1	22.2
6						60.7	47.9	40.0	34.5	30.4
7							65.2	52.9	45.0	39.3
8								68.8	57.1	49.3
9									71.7	60.6
10										74.1

5 per cent ranks

Rank order	Sample size 11	12	13	14	15	16	17	18	19	20
1	0.5	0.4	0.4	0.4	0.3	0.3	0.3	0.3	0.3	0.3
2	3.3	3.0	2.8	2.6	2.4	2.3	2.1	2.0	1.9	1.8
3	7.9	7.2	6.6	6.1	5.7	5.3	5.0	4.7	4.4	4.2
4	13.5	12.3	11.3	10.4	9.7	9.0	8.5	8.0	7.5	7.1
5	20.0	18.1	16.6	15.3	14.2	13.2	12.4	11.6	11.0	10.4
6	27.1	24.5	22.4	20.6	19.1	17.8	16.6	15.6	14.7	14.0
7	35.0	31.5	28.7	26.4	24.4	22.7	21.2	19.9	18.8	17.7
8	43.6	39.1	35.5	32.5	30.0	27.9	26.0	24.4	23.0	21.7
9	53.0	47.3	42.7	39.0	36.0	33.3	31.1	29.1	27.4	25.9
10	63.6	56.2	50.5	46.0	42.3	39.1	36.4	34.1	32.0	30.2
11	76.2	66.1	59.0	53.4	48.9	45.2	42.0	39.2	36.8	34.7
12		77.9	68.4	61.5	56.0	51.6	47.8	44.6	41.8	39.4
13			79.4	70.3	63.7	58.3	53.9	50.2	47.0	44.2
14				80.7	72.1	65.6	60.4	56.1	52.4	49.2
15					81.9	73.6	67.4	62.3	58.1	54.4
16						82.9	75.0	69.0	64.1	59.9
17							83.8	76.2	70.4	65.6
18								84.7	77.4	71.7
19									85.4	78.4
20										86.1

5 per cent ranks

Rank order	Sample size 21	22	23	24	25	26	27	28	29	30
1	0.2	0.2	0.2	0.2	0.2	0.2	0.2	0.2	0.2	0.2
2	1.7	1.6	1.6	1.5	1.4	1.4	1.3	1.3	1.2	1.2
3	4.0	3.8	3.7	3.5	3.4	3.2	3.1	3.0	2.9	2.8
4	6.8	6.5	6.2	5.9	5.7	5.4	5.2	5.0	4.9	4.7
5	9.9	9.5	9.0	8.6	8.2	7.9	7.6	7.3	7.0	6.8
6	13.2	12.6	12.0	11.5	11.0	10.6	10.1	9.8	9.4	9.1
7	16.8	16.0	15.2	14.6	13.9	13.4	12.9	12.4	11.9	11.5
8	20.6	19.6	18.6	17.8	17.0	16.3	15.7	15.1	14.5	14.0
9	24.5	23.3	22.2	21.2	20.2	19.4	18.6	17.9	17.2	16.6
10	28.6	27.1	25.8	24.6	23.6	22.6	21.7	20.8	20.1	19.3
11	32.8	31.1	29.6	28.2	27.0	25.8	24.8	23.8	22.9	22.1
12	37.2	35.3	33.5	31.9	30.5	29.2	28.0	26.9	25.9	25.0
13	41.7	39.5	37.5	35.8	34.1	32.7	31.3	30.1	28.9	27.9
14	46.4	43.9	41.7	39.7	37.9	36.2	34.7	33.3	32.0	30.8
15	51.3	48.5	46.0	43.7	41.7	39.8	38.2	36.6	35.2	33.9
16	56.3	53.2	50.4	47.9	45.6	43.6	41.7	40.0	38.4	37.0
17	61.6	58.0	54.9	52.1	49.6	47.4	45.3	43.5	41.7	40.2
18	67.1	63.1	59.6	56.5	53.8	51.3	49.1	47.0	45.1	43.4
19	72.9	68.4	64.5	61.1	58.0	55.3	52.9	50.6	48.6	46.7
20	79.3	74.1	69.6	65.8	62.5	59.5	56.8	54.3	52.1	50.1
21	86.7	80.2	75.1	70.8	67.0	63.7	60.8	58.1	55.7	53.5
22		87.3	81.0	76.0	71.8	68.2	64.9	62.0	59.4	57.0
23			87.8	81.7	76.9	72.8	69.2	66.1	63.2	60.6
24				88.3	82.4	77.7	73.7	70.2	67.1	64.3
25					88.7	83.0	78.5	74.6	71.2	68.1
26						89.2	83.6	79.2	75.4	72.0
27							89.5	84.1	79.8	76.1
28								89.9	84.7	80.5
29									90.2	85.1
30										90.5

5 per cent ranks

Rank order	Sample size 31	32	33	34	35	36	37	38	39	40
1	0.2	0.2	0.2	0.2	0.1	0.1	0.1	0.1	0.1	0.1
2	1.2	1.1	1.1	1.1	1.0	1.0	1.0	0.9	0.9	0.9
3	2.7	2.6	2.5	2.4	2.4	2.3	2.2	2.2	2.1	2.1
4	4.5	4.4	4.2	4.1	4.0	3.9	3.8	3.7	3.6	3.5
5	6.6	6.4	6.2	6.0	5.8	5.6	5.5	5.3	5.2	5.1
6	8.8	8.5	8.2	8.0	7.7	7.5	7.3	7.1	6.9	6.7
7	11.1	10.7	10.4	10.1	9.8	9.5	9.2	9.0	8.7	8.5
8	13.5	13.1	12.7	12.3	11.9	11.6	11.2	10.9	10.6	10.4
9	16.1	15.5	15.0	14.6	14.1	13.7	13.3	13.0	12.6	12.3
10	18.7	18.0	17.5	16.9	16.4	15.9	15.5	15.0	14.6	14.2
11	21.3	20.6	19.9	19.3	18.7	18.2	17.7	17.2	16.7	16.3
12	24.1	23.3	22.5	21.8	21.1	20.5	19.9	19.3	18.8	18.3
13	26.9	26.0	25.1	24.3	23.6	22.9	22.2	21.6	21.0	20.4
14	29.7	28.7	27.8	26.9	26.0	25.3	24.5	23.8	23.2	22.6
15	32.7	31.5	30.5	29.5	28.6	27.7	26.9	26.1	25.4	24.7
16	35.7	34.4	33.3	32.2	31.2	30.2	29.3	28.5	27.7	26.9
17	38.7	37.3	36.1	34.9	33.8	32.8	31.8	30.9	30.0	29.2
18	41.8	40.3	38.9	37.7	36.5	35.3	34.3	33.3	32.3	31.5
19	45.0	43.3	41.9	40.5	39.2	38.0	36.8	35.7	34.7	33.8
20	48.2	46.4	44.8	43.3	41.9	40.6	39.4	38.2	37.1	36.1
21	51.5	49.6	47.8	46.2	44.7	43.3	42.0	40.7	39.6	38.5
22	54.8	52.8	50.9	49.2	47.6	46.0	44.6	43.3	42.1	40.9
23	58.2	56.1	54.0	52.2	50.4	48.8	47.3	45.9	44.6	43.3
24	61.7	59.4	57.2	55.2	53.4	51.7	50.0	48.5	47.1	45.8
25	65.3	62.8	60.5	58.4	56.4	54.5	52.8	51.2	49.7	48.3
26	69.0	66.3	63.8	61.5	59.4	57.5	55.6	53.9	52.3	50.8
27	72.9	69.9	67.2	64.8	62.5	60.4	58.5	56.7	55.0	53.4
28	76.9	73.6	70.7	68.1	65.7	63.5	61.4	59.5	57.7	56.0
29	81.1	77.5	74.4	71.5	68.9	66.6	64.4	62.3	60.4	58.6
30	85.6	81.6	78.2	75.1	72.3	69.7	67.4	65.2	63.2	61.3
31	90.8	86.0	82.1	78.7	75.7	73.0	70.5	68.2	66.0	64.0
32		91.1	86.4	82.6	79.3	76.4	73.7	71.2	68.9	66.8
33			91.3	86.8	83.1	79.8	76.9	74.3	71.9	69.6
34				91.6	87.2	83.5	80.4	77.5	74.9	72.5
35					91.8	87.5	83.9	80.8	78.0	75.5
36						92.0	87.8	84.3	81.3	78.6
37							92.2	88.1	84.7	81.7
38								92.4	88.4	85.1
39									92.6	88.7
40										92.8

5 per cent ranks

	Sample size									
Rank order	41	42	43	44	45	46	47	48	49	50
1	0.1	0.1	0.1	0.1	0.1	0.1	0.1	0.1	0.1	0.1
2	0.9	0.9	0.8	0.8	0.8	0.8	0.8	0.7	0.7	0.7
3	2.0	2.0	1.9	1.9	1.8	1.8	1.8	1.7	1.7	1.7
4	3.4	3.3	3.2	3.2	3.1	3.0	3.0	2.9	2.8	2.8
5	4.9	4.8	4.7	4.6	4.5	4.4	4.3	4.2	4.1	4.0
6	6.6	6.4	6.3	6.1	6.0	5.8	5.7	5.6	5.5	5.4
7	8.3	8.1	7.9	7.7	7.5	7.4	7.2	7.1	6.9	6.8
8	10.1	9.8	9.6	9.4	9.2	9.0	8.8	8.6	8.4	8.2
9	12.0	11.7	11.4	11.1	10.9	10.6	10.4	10.1	9.9	9.7
10	13.9	13.5	13.2	12.9	12.6	12.3	12.0	11.8	11.5	11.3
11	15.8	15.4	15.1	14.7	14.4	14.0	13.7	13.4	13.1	12.9
12	17.8	17.4	17.0	16.6	16.2	15.8	15.4	15.1	14.8	14.5
13	19.9	19.4	18.9	18.4	18.0	17.6	17.2	16.8	16.5	16.1
14	22.0	21.4	20.9	20.4	19.9	19.4	19.0	18.6	18.2	17.8
15	24.1	23.5	22.9	22.3	21.8	21.3	20.8	20.4	19.9	19.5
16	26.2	25.6	24.9	24.3	23.7	23.2	22.7	22.2	21.7	21.2
17	28.4	27.7	27.0	26.3	25.7	25.1	24.5	24.0	23.5	23.0
18	30.6	29.8	29.1	28.4	27.7	27.0	26.4	25.8	25.3	24.7
19	32.9	32.0	31.2	30.4	29.7	29.0	28.3	27.7	27.1	26.5
20	35.1	34.2	33.3	32.5	31.7	31.0	30.3	29.6	28.9	28.3
21	37.4	36.5	35.5	34.6	33.8	33.0	32.2	31.5	30.8	30.1
22	39.8	38.7	37.7	36.8	35.9	35.0	34.2	33.4	32.7	32.0
23	42.1	41.0	39.9	38.9	38.0	37.1	36.2	35.4	34.6	33.8
24	44.5	43.3	42.2	41.1	40.1	39.2	38.2	37.4	36.5	35.7
25	46.9	45.7	44.5	43.3	42.3	41.3	40.3	39.4	38.5	37.6
26	49.4	48.1	46.8	45.6	44.5	43.4	42.3	41.4	40.4	39.5
27	51.9	50.5	49.1	47.9	46.7	43.5	44.4	43.4	42.4	41.5
28	54.4	52.9	51.5	50.1	48.9	47.7	46.5	45.5	44.4	43.4
29	56.9	55.4	53.9	52.5	51.1	49.9	48.7	47.5	46.4	45.4
30	59.5	57.9	56.3	54.8	53.4	52.1	50.8	49.6	48.5	47.4
31	62.1	60.4	58.7	57.2	55.7	54.3	53.0	51.7	50.5	49.4
32	64.8	63.0	61.2	59.6	58.0	56.6	55.2	53.9	52.6	51.4
33	67.5	65.6	63.8	62.0	60.4	58.9	57.4	56.0	54.7	53.5
34	70.3	68.2	66.3	64.5	62.8	61.2	59.7	58.2	56.8	55.5
35	73.1	71.0	68.9	67.0	65.2	63.5	61.9	60.4	59.0	57.6
36	76.1	73.7	71.6	69.6	67.7	65.9	64.2	62.7	61.2	59.7
37	79.0	76.6	74.3	72.2	70.2	68.3	66.6	64.9	63.4	61.9
38	82.2	79.5	77.1	74.8	72.8	70.8	69.0	67.2	65.6	64.0
39	85.4	82.6	80.0	77.6	75.4	73.3	71.4	69.6	67.8	66.2
40	88.9	85.8	82.9	80.4	78.0	75.9	73.8	71.9	70.1	68.4
41	93.0	89.2	86.1	83.3	80.8	78.5	76.4	74.3	72.5	70.7
42		93.1	89.4	86.4	83.7	81.2	78.9	76.8	74.8	73.0
43			93.3	89.7	86.7	84.0	81.6	79.3	77.3	75.3
44				93.4	89.9	86.9	84.3	81.9	79.7	77.7
45					93.6	90.1	87.2	84.6	82.3	80.1
46						93.7	90.3	87.5	84.9	82.6
47							93.8	90.5	87.7	85.2
48								94.0	90.7	87.9
49									94.1	90.9
50										94.2

95 per cent ranks

Rank order	*Sample size* 1	2	3	4	5	6	7	8	9	10
1	95.0	77.6	63.2	52.7	45.1	39.3	34.8	31.2	28.3	25.9
2		97.5	86.5	75.1	65.7	58.2	52.1	47.1	42.9	39.4
3			98.3	90.2	81.1	72.9	65.9	60.0	55.0	50.7
4				98.7	92.4	84.7	77.5	71.1	65.5	60.7
5					99.0	93.7	87.1	80.7	74.9	69.6
6						99.1	94.7	88.9	83.1	77.8
7							99.3	95.4	90.2	85.0
8								99.4	95.9	91.3
9									99.4	96.3
10										99.5

95 per cent ranks

Rank order	*Sample size* 11	12	13	14	15	16	17	18	19	20
1	23.8	22.1	20.6	19.3	18.1	17.1	16.2	15.3	14.6	13.9
2	36.4	33.9	31.6	29.7	27.9	26.4	25.0	23.8	22.6	21.6
3	47.0	43.8	41.0	38.5	36.3	34.4	32.6	31.0	29.6	28.3
4	56.4	52.7	49.5	46.6	44.0	41.7	39.6	37.7	35.9	34.4
5	65.0	60.9	57.3	54.0	51.1	48.4	46.1	43.9	41.9	40.1
6	72.9	68.5	64.5	61.0	57.7	54.8	52.2	49.8	47.6	45.6
7	80.0	75.5	71.3	67.5	64.0	60.9	58.0	55.4	53.0	50.8
8	86.5	81.9	77.6	73.6	70.0	66.7	63.6	60.8	58.2	55.8
9	92.1	87.7	83.4	79.4	75.6	72.1	68.9	65.9	63.2	60.6
10	96.7	92.8	88.7	84.7	80.9	77.3	74.0	70.9	68.0	65.3
11	99.5	97.0	93.4	89.6	85.8	82.2	78.8	75.6	72.6	69.8
12		99.6	97.2	93.9	90.3	86.8	83.4	80.1	77.0	74.1
13			99.6	97.4	94.3	91.0	87.6	84.4	81.3	78.3
14				99.6	97.6	94.7	91.5	88.4	85.3	82.3
15					99.7	97.7	95.0	92.0	89.0	86.0
16						99.7	97.9	95.3	92.5	89.6
17							99.7	98.0	95.6	92.9
18								99.7	98.1	95.8
19									99.7	98.2
20										99.7

95 per cent ranks

Rank order	*Sample size* 21	22	23	24	25	26	27	28	29	30
1	13.3	12.7	12.2	11.7	11.3	10.9	10.5	10.1	9.8	9.5
2	20.7	19.8	19.0	18.3	17.6	17.0	16.4	15.9	15.3	14.9
3	27.1	25.9	24.9	24.0	23.1	22.3	21.5	20.8	20.2	19.5
4	32.9	31.6	30.4	29.2	28.2	27.2	26.3	25.4	24.6	23.9
5	38.4	36.9	35.5	34.2	33.0	31.8	30.8	29.8	28.8	28.0
6	43.7	42.0	40.4	38.9	37.5	36.3	35.1	33.9	32.9	31.9
7	48.7	46.8	45.1	43.5	42.0	40.5	39.2	38.0	36.8	35.7
8	53.6	51.5	49.6	47.9	46.2	44.7	43.2	41.9	40.6	39.4
9	58.3	56.1	54.0	52.1	50.4	48.7	47.1	45.7	44.3	43.0
10	62.8	60.5	58.3	56.3	54.4	52.6	50.9	49.4	47.9	46.5
11	67.2	64.7	62.5	60.3	58.3	56.4	54.7	53.0	51.4	49.9
12	71.4	68.9	66.5	64.2	62.1	60.2	58.3	56.5	54.9	53.3
13	75.5	72.9	70.4	68.1	65.9	63.8	61.8	60.0	58.3	56.6
14	79.4	76.7	74.2	71.8	69.5	67.3	65.3	63.4	61.6	59.8
15	83.2	80.4	77.8	75.4	73.0	70.8	68.7	66.7	64.8	63.0
16	86.8	84.0	81.4	78.8	76.4	74.2	72.0	69.9	68.0	66.1
17	90.1	87.4	84.8	82.2	79.8	77.4	75.2	73.1	71.1	69.2
18	93.2	90.6	88.0	85.4	83.0	80.6	78.3	76.2	74.1	72.1
19	96.0	93.5	91.0	88.5	86.1	83.7	81.4	79.2	77.1	75.0
20	98.3	96.2	93.8	91.4	89.0	86.6	84.3	82.1	80.0	77.9
21	99.8	98.4	96.3	94.1	91.8	89.4	87.1	84.9	82.8	80.7
22		99.8	98.4	96.5	94.3	92.1	89.9	87.6	85.5	83.4
23			99.8	98.5	96.6	94.6	92.4	90.2	88.1	86.0
24				99.8	98.6	96.8	94.8	92.7	90.6	88.5
25					99.8	98.6	96.9	95.0	93.0	90.9
26						99.8	98.7	97.0	95.1	93.2
27							99.8	98.7	97.1	95.3
28								99.8	98.8	97.2
29									99.8	98.8
30										99.8

95 per cent ranks

	Sample size									
Rank order	31	32	33	34	35	36	37	38	39	40
1	9.2	8.9	8.7	8.4	8.2	8.0	7.8	7.6	7.4	7.2
2	14.4	14.0	13.6	13.2	12.9	12.5	12.2	11.9	11.6	11.3
3	18.9	18.4	17.9	17.4	16.9	16.5	16.1	15.7	15.3	14.9
4	23.2	22.5	21.9	21.3	20.7	20.2	19.6	19.2	18.7	18.3
5	27.1	26.4	25.6	24.9	24.3	23.6	23.1	22.5	22.0	21.4
6	31.0	30.1	29.3	28.5	27.7	27.0	26.3	25.7	25.1	24.5
7	34.7	33.7	32.8	31.9	31.1	30.3	29.5	28.8	28.1	27.5
8	38.3	37.2	36.2	35.2	34.3	33.4	32.6	31.8	31.1	30.4
9	41.8	40.6	39.5	38.5	37.5	36.5	35.6	34.8	34.0	33.2
10	45.2	43.9	42.8	41.6	40.6	39.6	38.6	37.7	36.8	36.0
11	48.5	47.2	46.0	44.8	43.6	42.5	41.5	40.5	39.6	38.7
12	51.8	50.4	49.1	47.8	46.6	45.5	44.4	43.3	42.3	41.4
13	55.0	53.6	52.2	50.8	49.6	48.3	47.2	46.1	45.0	44.0
14	58.2	56.7	55.2	53.8	52.4	51.2	50.0	48.8	47.7	46.6
15	61.3	59.7	58.1	56.7	55.3	54.0	52.7	51.5	50.3	49.2
16	64.3	62.7	61.1	59.5	58.1	56.7	55.4	54.1	52.9	51.7
17	67.3	65.6	63.9	62.3	60.8	59.4	58.0	56.7	55.4	54.2
18	70.3	68.5	66.7	65.1	63.5	62.0	60.6	59.3	57.9	56.7
19	73.1	71.3	69.5	67.8	66.2	64.7	63.2	61.8	60.4	59.1
20	75.9	74.0	72.2	70.5	68.8	67.2	65.7	64.3	62.9	61.5
21	78.7	76.7	74.9	73.1	71.4	69.8	68.2	66.7	65.3	63.9
22	81.3	79.4	77.5	75.7	74.0	72.3	70.7	69.1	67.7	66.2
23	83.9	82.0	80.1	78.2	76.4	74.7	73.1	71.5	70.0	68.5
24	86.5	84.5	82.5	80.7	78.9	77.1	75.5	73.9	72.3	70.8
25	88.9	86.9	85.0	83.1	81.3	79.5	77.8	76.2	74.6	73.1
26	91.2	89.3	87.3	85.4	83.6	81.8	80.1	78.4	76.8	75.3
27	93.4	91.5	89.6	87.7	85.9	84.1	82.3	80.7	79.0	77.4
28	95.5	93.6	91.8	89.9	88.1	86.3	84.5	82.8	81.2	79.6
29	97.3	95.6	93.8	92.0	90.2	88.4	86.7	85.0	83.3	81.7
30	98.8	97.4	95.8	94.0	92.3	90.5	88.8	87.1	85.4	83.7
31	99.8	98.9	97.5	95.9	94.2	92.5	90.8	89.1	87.4	85.8
32		99.8	98.9	97.6	96.0	94.4	92.7	91.0	89.4	87.7
33			99.8	98.9	97.6	96.1	94.5	92.9	91.3	89.6
34				99.8	99.0	97.7	96.2	94.7	93.1	91.5
35					99.9	99.0	97.8	96.3	94.8	93.3
36						99.9	99.0	97.8	96.4	94.9
37							99.9	99.1	97.9	96.5
38								99.9	99.1	97.9
39									99.9	99.1
40										99.9

95 per cent ranks

	Sample size									
Rank order	41	42	43	44	45	46	47	48	49	50
1	7.0	6.9	6.7	6.6	6.4	6.3	6.2	6.1	5.9	5.8
2	11.1	10.8	10.6	10.3	10.1	9.9	9.7	9.5	9.3	9.1
3	14.6	14.2	13.9	13.6	13.3	13.1	12.8	12.5	12.3	12.1
4	17.8	17.4	17.1	16.7	16.3	16.0	15.7	15.4	15.1	14.8
5	21.0	20.5	20.0	19.6	19.2	18.8	18.4	18.1	17.7	17.4
6	23.9	23.4	22.9	22.4	22.0	21.5	21.1	20.7	20.3	19.9
7	26.9	26.3	25.7	25.2	24.6	24.1	23.7	23.2	22.7	22.3
8	29.7	29.0	28.4	27.6	27.2	26.7	26.2	25.7	25.2	24.7
9	32.5	31.8	31.1	30.4	29.8	29.2	28.6	28.1	27.5	27.0
10	35.2	34.4	33.7	33.0	32.3	31.7	31.0	30.4	29.9	29.3
11	37.9	37.0	36.2	35.5	34.8	34.1	33.4	32.8	32.2	31.6
12	40.5	39.6	38.8	38.0	37.2	36.5	35.8	35.1	34.4	33.8
13	43.1	42.1	41.3	40.4	39.6	38.8	38.1	37.3	36.6	36.0
14	45.6	44.6	43.7	42.8	42.0	41.1	40.3	39.6	38.8	38.1
15	48.1	47.1	46.1	45.2	44.3	43.4	42.6	41.8	41.0	40.3
16	50.6	49.5	48.5	47.5	46.6	45.7	44.8	44.0	43.2	42.4
17	53.1	52.0	50.9	49.9	48.9	47.9	47.0	46.1	45.3	44.5
18	55.5	54.3	53.2	52.1	51.1	50.1	49.2	48.3	47.4	46.5
19	57.9	56.7	55.5	54.4	53.3	52.3	51.3	50.4	49.5	48.6
20	60.2	59.0	57.8	56.7	55.5	54.5	53.5	52.5	51.5	50.6
21	62.6	61.3	60.1	58.9	57.7	56.6	55.6	54.5	53.6	52.6
22	64.9	63.5	62.3	61.1	59.9	58.7	57.7	56.6	55.6	54.6
23	67.1	65.8	64.5	63.2	62.0	60.8	59.7	58.6	57.6	56.6
24	69.4	68.0	66.7	65.4	64.1	62.9	61.8	60.6	59.6	58.5
25	71.6	70.2	68.8	67.5	66.2	65.0	63.8	62.6	61.5	60.5
26	73.8	72.3	70.9	69.6	68.3	67.0	65.8	64.6	63.5	62.4
27	75.9	74.4	73.0	71.6	70.3	69.0	67.8	66.6	65.4	64.3
28	78.0	76.5	75.1	73.7	72.3	71.0	69.7	68.5	67.3	66.2
29	80.1	78.6	77.1	75.7	74.3	73.0	71.7	70.4	69.2	68.0
30	82.2	80.6	79.1	77.7	76.3	74.9	73.6	72.3	71.1	69.9
31	84.2	82.6	81.1	79.6	78.2	76.8	75.5	74.2	72.9	71.7
32	86.1	84.6	83.0	81.6	80.1	78.7	77.3	76.0	74.7	73.5
33	88.0	86.5	84.9	83.4	82.0	80.6	79.2	77.8	76.5	75.3
34	89.9	88.3	86.8	85.3	83.8	82.4	81.0	79.7	78.3	77.0
35	91.7	90.2	88.6	87.1	85.6	84.2	82.8	81.4	80.1	78.8
36	93.4	91.9	90.4	88.9	87.4	86.0	84.6	83.2	81.8	80.5
37	95.1	93.6	92.1	90.5	89.2	87.7	86.3	84.9	83.5	82.2
38	96.6	95.2	93.7	92.3	90.8	89.4	88.0	86.6	85.2	83.9
39	98.0	96.7	95.3	93.9	92.5	91.0	89.6	88.2	86.9	85.5
40	99.1	98.0	96.8	95.4	94.0	92.6	91.2	89.9	88.5	87.1
41	99.9	99.1	98.1	96.8	95.5	94.2	92.8	91.4	90.1	88.7
42		99.9	99.2	98.1	96.9	95.6	94.3	93.0	91.6	90.3
43			99.9	99.2	98.2	97.0	95.7	94.4	93.1	91.8
44				99.9	99.2	98.2	97.0	95.8	94.5	93.2
45					99.9	99.2	98.2	97.1	95.9	94.6
46						99.9	99.2	98.3	97.2	96.0
47							99.9	99.3	98.3	97.2
48								99.9	99.3	98.3
49									99.9	99.3
50										99.9

Median ranks

Rank order	Sample size 1	2	3	4	5	6	7	8	9	10
1	50.0	29.3	20.6	15.9	12.9	10.9	9.4	8.3	7.4	6.7
2		70.7	50.0	38.6	31.4	26.4	22.8	20.1	18.0	16.2
3			79.4	61.4	50.0	42.1	36.4	32.1	28.6	25.9
4				84.1	68.6	57.9	50.0	44.0	39.3	35.5
5					87.1	73.6	63.6	56.0	50.0	45.2
6						89.1	77.2	67.9	60.7	54.8
7							90.6	79.9	71.4	64.5
8								91.7	82.0	74.1
9									92.6	83.8
10										93.3

Median ranks

Rank order	Sample size 11	12	13	14	15	16	17	18	19	20
1	6.1	5.6	5.2	4.8	4.5	4.2	4.0	3.8	3.6	3.4
2	14.8	13.6	12.6	11.7	10.9	10.3	9.7	9.2	8.7	8.3
3	23.6	21.7	20.0	18.6	17.4	16.4	15.4	14.6	13.8	13.1
4	32.4	29.8	27.5	25.6	23.9	22.5	21.2	20.0	19.0	18.1
5	41.2	37.9	35.0	32.6	30.5	28.6	26.9	25.5	24.2	23.0
6	50.0	46.0	42.5	39.5	37.0	34.7	32.7	30.9	29.3	27.9
7	58.8	54.0	50.0	46.5	43.5	40.8	38.5	36.4	34.5	32.8
8	67.6	62.1	57.5	53.5	50.0	46.9	44.2	41.8	39.7	37.7
9	76.4	70.2	65.0	60.5	56.5	53.1	50.0	47.3	44.8	42.6
10	85.2	78.3	72.5	67.4	63.0	59.2	55.8	52.7	50.0	47.5
11	93.9	86.4	80.0	74.4	69.5	65.3	61.5	58.2	55.2	52.5
12		94.4	87.4	81.4	76.1	71.4	67.3	63.6	60.3	57.4
13			94.8	88.3	82.6	77.5	73.1	69.1	65.5	62.3
14				95.2	89.1	83.6	78.8	74.5	70.7	67.2
15					95.5	89.7	84.6	80.0	75.8	72.1
16						95.8	90.3	85.4	81.0	77.0
17							96.0	90.8	86.2	81.9
18								96.2	91.3	86.9
19									96.4	91.7
20										96.6

Median ranks

Rank order	*Sample size* 21	22	23	24	25	26	27	28	29	30
1	3.2	3.1	3.0	2.8	2.7	2.6	2.5	2.4	2.4	2.3
2	7.9	7.5	7.2	6.9	6.6	6.4	6.1	5.9	5.7	5.5
3	12.5	12.0	11.5	11.0	10.6	10.2	9.8	9.4	9.1	8.8
4	17.2	16.4	15.7	15.1	14.5	13.9	13.4	13.0	12.5	12.1
5	21.9	20.9	20.0	19.2	18.4	17.7	17.1	16.5	15.9	15.4
6	26.6	25.4	24.3	23.3	22.4	21.5	20.7	20.0	19.3	18.7
7	31.3	29.9	28.6	27.4	26.3	25.3	24.4	23.5	22.7	22.0
8	35.9	34.3	32.9	31.5	30.3	29.1	28.1	27.1	26.1	25.3
9	40.6	38.8	37.1	35.6	34.2	32.9	31.7	30.6	29.6	28.6
10	45.3	43.3	41.4	39.7	38.2	36.7	35.4	34.1	33.0	31.9
11	50.0	47.8	45.7	43.8	42.1	40.5	39.0	37.7	36.4	35.2
12	54.7	52.2	50.0	47.9	46.1	44.3	42.7	41.2	39.8	38.5
13	59.4	56.7	54.3	52.1	50.0	48.1	46.3	44.7	43.2	41.8
14	64.1	61.2	58.6	56.2	53.9	51.9	50.0	48.2	46.6	45.1
15	68.7	65.7	62.9	60.3	57.9	55.7	53.7	51.8	50.0	48.4
16	73.4	70.1	67.1	64.4	61.8	59.5	57.3	55.3	53.4	51.6
17	78.1	74.6	71.4	68.5	65.8	63.3	61.0	58.8	56.8	54.9
18	82.8	79.1	75.7	72.6	69.7	67.1	64.6	62.4	60.2	58.2
19	87.5	83.6	80.0	76.7	73.7	70.9	68.3	65.9	63.6	61.5
20	92.1	88.0	84.3	80.8	77.6	74.7	71.9	69.4	67.0	64.8
21	96.8	92.5	88.5	84.9	81.6	78.5	75.6	72.9	70.5	68.1
22		96.9	92.8	89.0	85.5	82.3	79.3	76.5	73.9	71.4
23			97.0	93.1	89.4	86.1	82.9	80.0	77.3	74.7
24				97.2	93.4	89.8	86.6	83.5	80.7	78.0
25					97.3	93.6	90.2	87.0	84.1	81.3
26						97.4	93.9	90.6	87.5	84.6
27							97.5	94.1	90.9	87.9
28								97.6	94.3	91.2
29									97.6	94.5
30										97.7

Median ranks

Rank order	*Sample size* 31	32	33	34	35	36	37	38	39	40
1	2.2	2.1	2.1	2.0	2.0	1.9	1.9	1.8	1.8	1.7
2	5.4	5.2	5.0	4.9	4.7	4.6	4.5	4.4	4.3	4.2
3	8.5	8.3	8.0	7.8	7.6	7.4	7.2	7.0	6.8	6.6
4	11.7	11.4	11.0	10.7	10.4	10.1	9.8	9.6	9.3	9.1
5	14.9	14.4	14.0	13.6	13.2	12.9	12.5	12.2	11.9	11.6
6	18.1	17.5	17.0	16.5	16.0	15.6	15.2	14.8	14.4	14.1
7	21.3	20.6	20.0	19.4	18.9	18.4	17.9	17.4	17.0	16.5
8	24.5	23.7	23.0	22.3	21.7	21.1	20.5	20.0	19.5	19.0
9	27.7	26.8	26.0	25.2	24.5	23.9	23.2	22.6	22.0	21.5
10	30.9	29.9	29.0	28.2	27.4	26.6	25.9	25.2	24.6	24.0
11	34.0	33.0	32.0	31.1	30.2	29.4	28.6	27.8	27.1	26.4
12	37.2	36.1	35.0	34.0	33.0	32.1	31.3	30.4	29.7	28.9
13	40.4	39.2	38.0	36.9	35.9	34.9	33.9	33.0	32.2	31.4
14	43.6	42.3	41.0	39.8	38.7	37.6	36.6	35.7	34.7	33.9
15	46.8	45.4	44.0	42.7	41.5	40.4	39.3	38.3	37.3	36.4
16	50.0	48.5	47.0	45.6	44.3	43.1	42.0	40.9	39.8	38.8
17	53.2	51.5	50.0	48.5	47.2	45.9	44.6	43.5	42.4	41.3
18	56.4	54.6	53.0	51.5	50.0	48.6	47.3	46.1	44.9	43.8
19	59.6	57.7	56.0	54.4	52.8	51.4	50.0	48.7	47.5	46.3
20	62.8	60.8	59.0	57.3	55.7	54.1	52.7	51.3	50.0	48.8
21	66.0	63.9	62.0	60.2	58.5	56.9	55.4	53.9	52.5	51.2
22	69.1	67.0	65.0	63.1	61.3	59.6	58.0	56.5	55.1	53.7
23	72.3	70.1	68.0	66.0	64.1	62.4	60.7	59.1	57.6	56.2
24	75.5	73.2	71.0	68.9	67.0	65.1	63.4	61.7	60.2	58.7
25	78.7	76.3	74.0	71.8	69.8	67.9	66.1	64.3	62.7	61.2
26	81.9	79.4	77.0	74.8	72.6	70.6	68.7	67.0	65.3	63.6
27	85.1	82.5	80.0	77.7	75.5	73.4	71.4	69.6	67.8	66.1
28	88.3	85.6	83.0	80.6	78.3	76.1	74.1	72.2	70.3	68.6
29	91.5	88.6	86.0	83.5	81.1	78.9	76.8	74.8	72.9	71.1
30	94.6	91.7	89.0	86.4	84.0	81.6	79.5	77.4	75.4	73.6
31	97.8	94.8	92.0	89.3	86.8	84.4	82.1	80.0	78.0	76.0
32		97.9	95.0	92.2	89.6	87.1	84.8	82.6	80.5	78.5
33			97.9	95.1	92.4	89.9	87.5	85.2	83.0	81.0
34				98.0	95.3	92.6	90.2	87.8	85.6	83.5
35					98.0	95.4	92.8	90.4	88.1	85.9
36						98.1	95.5	93.0	90.7	88.4
37							98.1	95.6	93.2	90.9
38								98.2	95.7	93.4
39									98.2	95.8
40										98.3

Median ranks

Rank order	*Sample size* 41	42	43	44	45	46	47	48	49	50
1	1.7	1.6	1.6	1.6	1.5	1.5	1.5	1.4	1.4	1.4
2	4.1	4.0	3.9	3.8	3.7	3.6	3.5	3.5	3.4	3.3
3	6.5	6.3	6.2	6.0	5.9	5.8	5.6	5.5	5.4	5.3
4	8.9	8.7	8.5	8.3	8.1	7.9	7.8	7.6	7.4	7.3
5	11.3	11.0	10.8	10.5	10.3	10.1	9.9	9.7	9.5	9.3
6	13.7	13.4	13.1	12.8	12.5	12.2	12.0	11.7	11.5	11.3
7	16.1	15.8	15.4	15.0	14.7	14.4	14.1	13.8	13.5	13.3
8	18.6	18.1	17.7	17.3	16.9	16.6	16.2	15.9	15.5	15.2
9	21.0	20.5	20.0	19.6	19.1	18.7	18.3	17.9	17.6	17.2
10	23.4	22.8	22.3	21.8	21.3	20.9	20.4	20.0	19.6	19.2
11	25.8	25.2	24.6	24.1	23.5	23.0	22.5	22.1	21.6	21.2
12	28.2	27.6	26.9	26.3	25.7	25.2	24.7	24.1	23.7	23.2
13	30.6	29.9	29.2	28.6	27.9	27.3	26.8	26.2	25.7	25.2
14	33.1	32.3	31.5	30.8	30.1	29.5	28.9	28.3	27.7	27.2
15	35.5	34.6	33.8	33.1	32.4	31.7	31.0	30.3	29.7	29.1
16	37.9	37.0	36.2	35.3	34.6	33.8	33.1	32.4	31.8	31.1
17	40.3	39.4	38.5	37.6	36.8	36.0	35.2	34.5	33.8	33.1
18	42.7	41.7	40.8	39.9	39.0	38.1	37.3	36.6	35.8	35.1
19	45.2	44.1	43.1	42.1	41.2	40.3	39.4	38.6	37.8	37.1
20	47.6	46.5	45.4	44.4	43.4	42.4	41.6	40.7	39.9	39.1
21	50.0	48.8	47.7	46.6	45.6	44.6	43.7	42.8	41.9	41.1
22	52.4	51.2	50.0	48.9	47.8	46.8	45.8	44.8	43.9	43.0
23	54.8	53.5	52.3	51.1	50.0	48.9	47.9	46.9	45.9	45.0
24	57.3	55.9	54.6	53.4	52.2	51.1	50.0	49.0	48.0	47.0
25	59.7	58.3	56.9	55.6	54.4	53.2	52.1	51.0	50.0	49.0
26	62.1	60.6	59.2	57.9	56.6	55.4	54.2	53.1	52.0	51.0
27	64.5	63.0	61.5	60.1	58.8	57.6	56.3	55.2	54.1	53.0
28	66.9	65.4	63.8	62.4	61.0	59.7	58.5	57.2	56.1	55.0
29	69.4	67.7	66.2	64.7	63.2	61.9	60.6	59.3	58.1	57.0
30	71.8	70.1	68.5	66.9	65.4	64.0	62.7	61.4	60.1	58.9
31	74.2	72.4	70.8	69.2	67.6	66.2	64.8	63.4	62.2	60.9
32	76.6	74.8	73.1	71.4	69.9	68.3	66.9	65.5	64.2	62.9
33	79.0	77.2	75.4	73.7	72.1	70.5	69.0	67.6	66.2	64.9
34	81.4	79.5	77.7	75.9	74.3	72.7	71.1	69.7	68.2	66.9
35	83.9	81.9	80.0	78.2	76.5	74.8	73.2	71.7	70.3	68.9
36	86.3	84.2	82.3	80.4	78.7	77.0	75.3	73.8	72.3	70.9
37	88.7	86.6	84.6	82.7	80.9	79.1	77.5	75.9	74.3	72.8
38	91.1	89.0	86.9	85.0	83.1	81.3	79.6	77.9	76.3	74.8
39	93.5	91.3	89.2	87.2	85.3	83.4	81.7	80.0	78.4	76.8
40	95.9	93.7	91.5	89.5	87.5	85.6	83.8	82.1	80.4	78.8
41	98.3	96.0	93.8	91.7	89.7	87.8	85.9	84.1	82.4	80.8
42		98.4	96.1	94.0	91.9	89.9	88.0	86.2	84.5	82.8
43			98.4	96.2	94.1	92.1	90.1	88.3	86.5	84.8
44				98.4	96.3	94.2	92.2	90.3	88.5	86.8
45					98.5	96.4	94.4	92.4	90.5	88.7
46						98.5	96.5	94.5	92.6	90.7
47							98.5	96.5	94.6	92.7
48								98.6	96.6	94.7
49									98.6	96.7
50										98.6

For sample sizes greater than 50, use

Mean rank = $[100i/(n+1)]$

Appendix 3: The Elements of Probability Theory

A3.1 Definition

'If an experiment can result in n mutually exclusive and equally likely possible outcomes, n_1 of which correspond to the occurrence of an event E_1 then the probability of the event E_1 is n_1/n or $\Pr(E_1) = n_1/n$' (Wine)

A3.2 Venn Diagram

A pictorial illustration of probabilities is shown in a Venn diagram, see Figure A3.1. Note that Venn diagrams can have areas which are proportional to the corresponding probabilities.

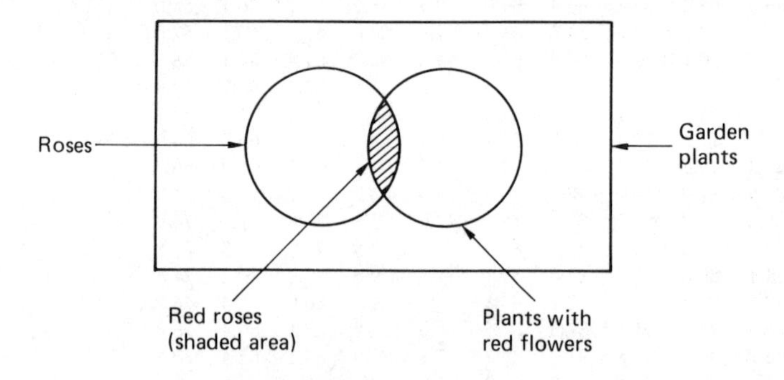

Figure A3.1 A Venn diagram

A3.3 Joint Probabilities (Intersection)

Probability of A and B $\Pr(A \text{ and } B)$: illustrated by the shaded area of Figure A3.1 (red roses).

A3.4 Addition of Probabilities (Union)

Probability of A or B Pr(A or B): illustrated by total encircled area of Figure A3.1 (plants with red flowers or roses).

A3.5 Mutually Exclusive Events

A and B are mutually exclusive if Pr(A and B) = 0. The probability that a person is (legally) both married and unmarried is zero.

A4.6 Complement of an Event

$\bar{A}$ ('not A' in probability theory) is the complement of A.

$$\Pr(\bar{A}) = 1 - \Pr(A)$$

A3.7 Independent Events

Two events R and U are defined as being independent if

$$\Pr(U \text{ and } R) = \Pr(U)\Pr(R)$$

Consider the case of a man who goes for a walk every day. Over a considerable period of time he assembles data on whether it rains or not during his walk. Neglecting variations between the seasons, he may then use his historical data to predict that there is a probability of 0.8 that it will rain on a particular day.

We can then write

$$\Pr(R) = 0.8$$

and using $\Pr(\bar{R})$ to denote the probability that it does not rain, then

$$\Pr(\bar{R}) = 0.2$$

since the events R and $\bar{R}$ are exhaustive and mutually exclusive.

Let the man toss a coin each day before his walk to decide whether or not he carries an umbrella. Then using the obvious notation,

$$\Pr(U) = 0.5$$

and

$$\Pr(\bar{U}) = 0.5$$

It is clear that the events R and U are independent, as are the pairs of events R and $\bar{U}$, $\bar{R}$ and U, $\bar{R}$ and $\bar{U}$. A Venn diagram illustrating the probabilities of these four joint events is as shown in Figure A3.2. The figure illustrates what we would expect from first principles, namely

$$\Pr(R \text{ and } U) = \Pr(R)\Pr(U) = 0.8 \times 0.5 = 0.4$$
$$\Pr(R \text{ and } \bar{U}) = \Pr(R)\Pr(\bar{U}) = 0.8 \times 0.5 = 0.4$$
$$\Pr(\bar{R} \text{ and } U) = \Pr(\bar{R})\Pr(U) = 0.2 \times 0.5 = 0.1$$
$$\Pr(\bar{R} \text{ and } \bar{U}) = \Pr(\bar{R})\Pr(\bar{U}) = 0.2 \times 0.5 = 0.1$$

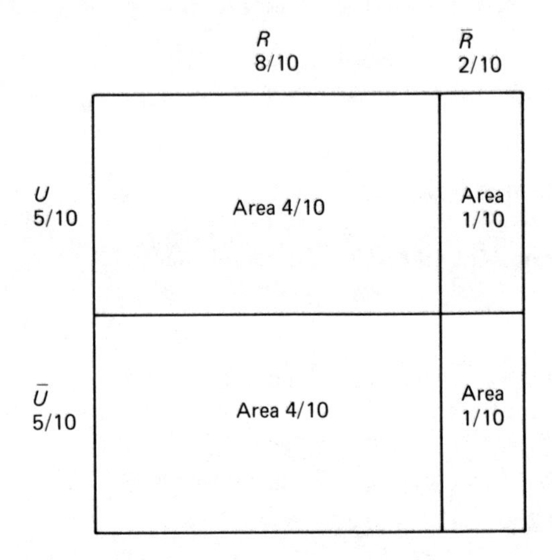

Figure A3.2 Venn diagram illustrating independent events

A3.8 Dependent Events – Conditional Probabilities

Consider the same man of section A3.7, who goes for a walk each day. Let him now, before going on his walk, look out of the window and use his judgement to decide whether or not to carry an umbrella. If we assume that his chance of making a correct choice is better than obtained by tossing a coin, then the events R and U, R and $\bar{U}$, $\bar{R}$ and U, $\bar{R}$ and $\bar{U}$ are no longer independent. It is necessary now to introduce conditional probabilities of the type $\Pr(U \mid R)$ which is the conditional probability of the event U given that the event R has occurred. If the man described above makes a correct choice 3 times out of 4 whenever it rains, then

$$\Pr(U \mid R) = 3/4$$

and correspondingly

$$\Pr(\bar{U} \mid R) = 1/4$$

If he makes a correct choice 7 times out of 10 whenever it does not rain, then

$$\Pr(\bar{U} \mid \bar{R}) = 7/10$$

and correspondingly

$$\Pr(U \mid \bar{R}) = 3/10$$

The Venn diagram illustrating the combinations of these two dependent events appears as shown in Figure A3.3.

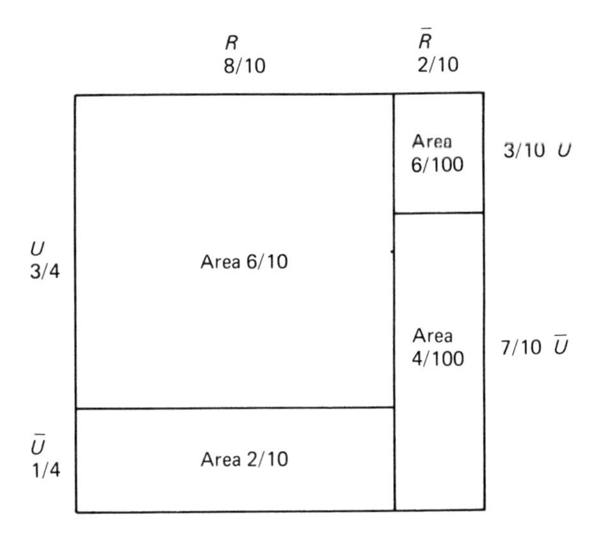

Figure A3.3 Venn diagram illustrating dependent events

It follows from the definition of conditional probabilities that

$$\Pr(U \mid R) = \frac{\Pr(U \text{ and } R)}{\Pr(R)}$$

so that

$$\Pr(U \text{ and } R) = \Pr(U \mid R)\Pr(R)$$

It can be seen from Figure A3.3 that

$$\Pr(U \text{ and } R) = \frac{3}{4} \times \frac{8}{10} = \frac{6}{10}$$

It should be noted that $\Pr(U \text{ and } R)$ does not equal $\Pr(U)\Pr(R)$, since

$$\Pr(U) = \frac{66}{100} \text{ and } \Pr(R) = \frac{8}{10}.$$

The data of Figure A3.3 consists essentially of four pieces:

$\Pr(U \text{ and } R) = 60/100$
$\Pr(U \text{ and } \bar{R}) = 6/100$
$\Pr(\bar{U} \text{ and } R) = 20/100$
$\Pr(\bar{U} \text{ and } \bar{R}) = 14/100$

These data can be re-assembled to yield the Venn diagram of Figure A3.4, which illustrates that it is possible to construct four other conditional probabilities, namely

$\Pr(R \mid U) = 10/11$
$\Pr(\bar{R} \mid U) = 1/11$
$\Pr(R \mid \bar{U}) = 10/17$
$\Pr(\bar{R} \mid \bar{U}) = 7/17$

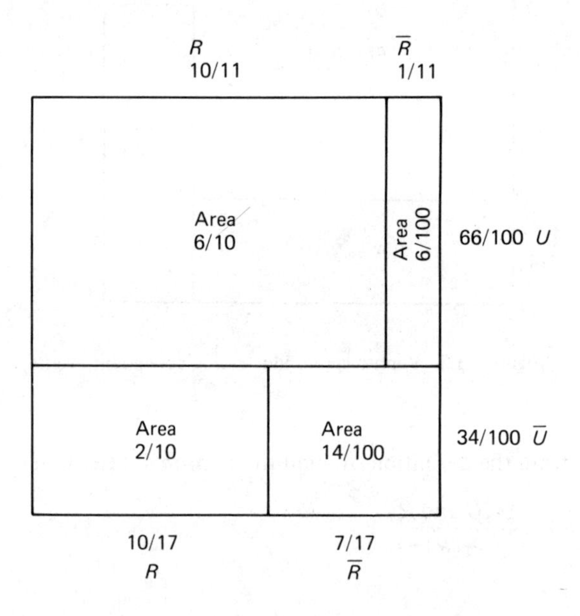

Figure A3.4 A re-arrangement of Figure A3.3

A3.9 Conditional Probabilities

$\Pr(A \mid B)$ = probability of event A given that event B has occurred.

As an example, consider an urn containing 2 red balls and 3 black balls. One ball is removed and then a second. The probability that the second ball is red clearly depends upon the colour of the first ball that was removed.

We can see that

$$\Pr(R_2 \mid R_1) = 1/4$$
$$\Pr(R_2 \mid B_1) = 2/4$$

A3.10 The Theorem of Total Probabilities

The Theorem of Total Probabilities states that

$$\Pr(S) = \Pr(S \mid E_1)\Pr(E_1) + \Pr(S \mid E_2)\Pr(E_2) + \ldots + \Pr(S \mid E_n)\Pr(E_n)$$

where $\Pr(S)$ is the probability of an event S

$\Pr(S \mid E_i)$ is the conditional probability of event S given that event E_i has occurred
$\Pr(E_i)$ is the probability of an event E_i, and it is necessary that the events $E_1, E_2, \ldots E_n$ are mutually exclusive (only one can happen) and exhaustive (one or other must happen) so that

$$\Pr(E_1) + \Pr(E_2) + \ldots + \Pr(E_n) = 1$$

The theorem is illustrated by Figure A3.5, in which the four events E_1, E_2, E_3 and E_4 are seen to be mutually exclusive and exhaustive. The shaded area shows the probability of the event S which can occur in conjunction with any of the events E_1, E_2, E_3 and E_4. The probability of S given that E_1 has occurred (the conditional probability $\Pr(S \mid E_1)$) enables the area of the top left-hand corner of the figure to be found, since this is $\Pr(S \mid E_1)\Pr(E_1)$. In a similar way, the left-hand shaded area for any event E_i is given by $\Pr(S \mid E_i)\Pr(E_i)$. Thus, the probability of the event S occurring is

$$\Pr(S) = \Pr(S \mid E_1)\Pr(E_1) + \Pr(S \mid E_2)\Pr(E_2) + \Pr(S \mid E_3)\Pr(E_3) + \Pr(S \mid E_4)\Pr(E_4)$$

It can readily be seen, by reference to the figure, that the events $E_1, E_2 \ldots$ must be exhaustive, since otherwise some of the shaded area might be omitted in the summation. If the events $E_1, E_2 \ldots$ are not mutually exclusive, then some of the areas of Figure A3.5 will overlap and the probabilities which they represent will be counted twice over. It should be noted that it is possible for any of the $\Pr(S \mid E_i)$ to be zero, although this is not shown in the figure.

As an illustration of the use of the Theorem of Total Probabilities, consider the withdrawal, without replacement, of two balls taken at random from an urn containing 2 red balls and 3 black balls. The question to be answered is: what is the probability that a red ball will be the second selected ball?

Let $R_i(B_i)$ be the event that a red (black) ball is removed in the ith position. Then it can readily be seen that the events R_1 and B_1 are mutually exclusive and exhaustive. Applying the theorem of total probabilities to find the probability of the event R_2 gives

$$\Pr(R_2) = \Pr(R_2 \mid R_1)\Pr(R_1) + \Pr(R_2 \mid B_1)\Pr(B_1)$$

This relationship is illustrated in Figure A3.6, in which the shaded area represents $\Pr(R_2)$.

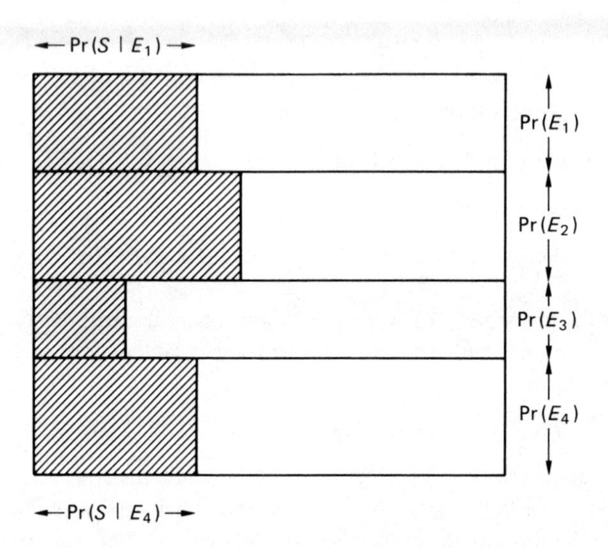

Figure A3.5 Venn diagram illustrating the Theorem of Total Probabilities

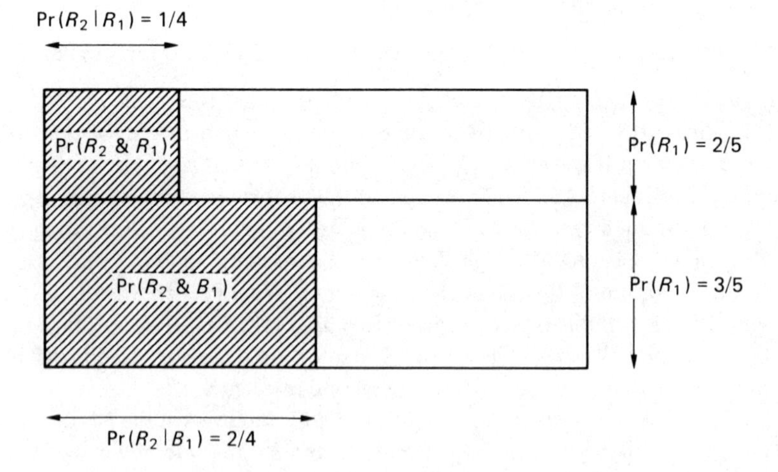

Figure A3.6 Venn diagram illustrating the withdrawal of two balls from an urn

It can readily be seen that

$$\Pr(R_1) = 2/5 \qquad \Pr(B_1) = 3/5$$
$$\Pr(R_2 \mid R_1) = 1/4 \qquad \Pr(R_2 \mid B_1) = 2/4$$

So that $\Pr(R_2) = 2/20 + 6/20 = 8/20$.

This can be readily checked by considering the red balls to be numbered 1 and 2, and the black balls to be numbered 3, 4 and 5, and then setting out all possible outcomes of the experiment. These are

12	21	31	41	51
13	23	32	42	52
14	24	34	43	53
15	25	35	45	54

It can be seen that the proportion of outcomes in which a red ball is drawn second (shown in box) is 8/20.

It can also be verified by examination of the first two columns that $\Pr(R_2 \mid R_1) = 2/8 = 1/4$, and by examination of that last three columns that $\Pr(R_2 \mid B_1) = 4/8 = 2/4$.

A3.11 Probability Tree Diagram

Probability tree diagrams may be used to show how the outcomes of successive trials may be combined to give various end events. The probabilities of the end events may be formed by multiplying the probabilities of the successive trials, so long as these probabilities are independent.

Consider a trial which consists of the successive tosses of a 'good' coin. The trial is halted when a head is thrown or after four successive tails. The possible outcomes of the successive trials are shown in Figure A3.7. It can be seen that there are five possible end events, namely

$$H_1, \quad T_1H_2, \quad T_1T_2H_3, \quad T_1T_2T_3H_4 \quad \text{and} \quad T_1T_2T_3T_4$$

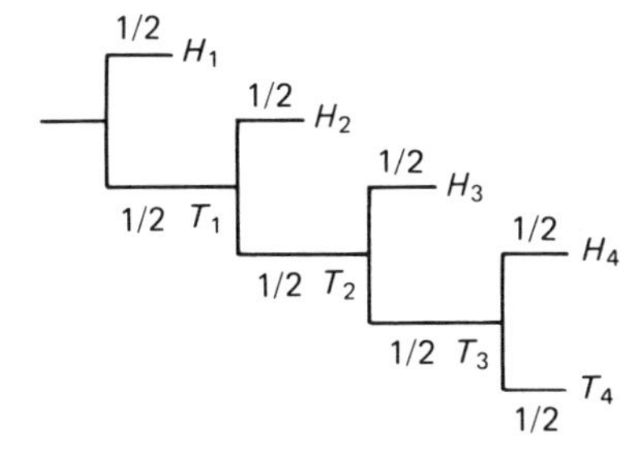

Figure A3.7 A probability tree diagram

The probabilities of these end events are given by

$$\begin{aligned}
&\Pr(H_1) &&= 1/2\\
&\Pr(T_1H_2) &&= 1/4\\
&\Pr(T_1T_2H_3) &&= 1/8\\
&\Pr(T_1T_2H_3H_4) &&= 1/16\\
&\Pr(T_1T_2T_3T_4) &&= 1/16\\
&\text{Total} &&= 1
\end{aligned}$$

It is permissible to multiply the successive probabilities to find the probability of the end events, since these probabilities are clearly independent.

Exercises

A3.1. Three people – A, B and C – fire at a target. The probability of scoring a bull is 0.3, 0.5 and 0.7 respectively. After each has fired once, it is found that only one person has scored a bull; what is the probability that the person was A?

A3.2. The three people – A, B and C – of Exercise **A3.1** take part in a trial. Person A fires first; if he scores a bull, the trial is terminated. If he misses, then B fires. If B scores a bull, the trial is terminated. If B misses, then C fires. What is the probability that the bull has been hit at the end of the trial?

Answers to Exercises

A3.1. It is reasonable to assume statistical independence between each person's attempt, so that all possible results of the trial can be obtained by the expansion of

$$(A+\bar{A})(B+\bar{B})(C+\bar{C}) = 1$$

where A represents the probability that person A scores a bull, and $\bar{A}$ that he misses, with similar nomenclature for B and C.

Only three terms, namely $A\,\bar{B}\,\bar{C}$, $\bar{A}\,B\,\bar{C}$ and $\bar{A}\,\bar{B}\,C$, represent these results with only one bull; their probabilities are

$$\begin{aligned}
&A\,\bar{B}\,\bar{C} &&= 0.3 \times 0.5 \times 0.3 = 0.045\\
&\bar{A}\,B\,\bar{C} &&= 0.7 \times 0.5 \times 0.3 = 0.105\\
&\bar{A}\,\bar{B}\,C &&= 0.7 \times 0.5 \times 0.7 = \underline{0.245}\\
&\text{Total} &&= \underline{0.395}
\end{aligned}$$

Probability that A hits the bull (given only one hit) is thus

$$0.045/0.395 = 9/79$$

A3.2. The results of this trial are not statistically independent. Possible outcomes are shown in Figure A3.8 from which it can be seen that the probability of a hit is

$$0.3 + 0.35 + 0.245 = 0.895$$

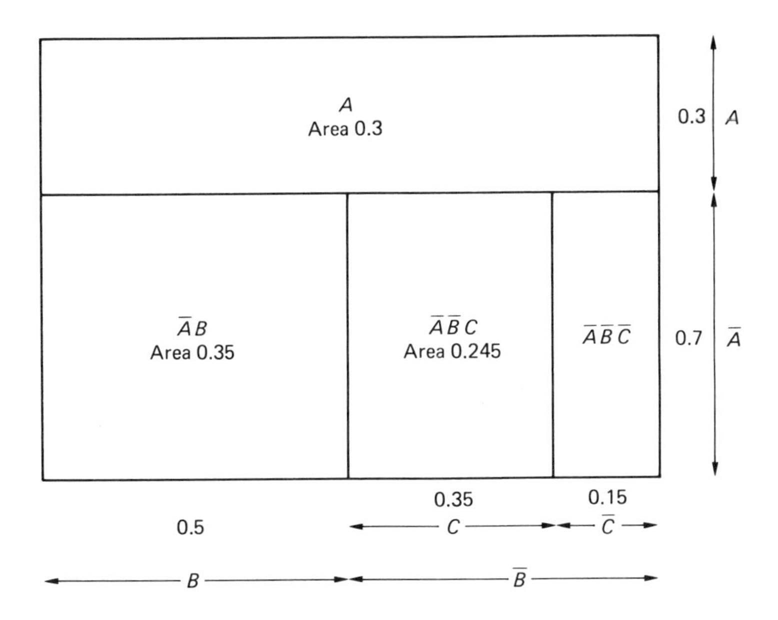

Figure A3.8 Venn diagram for Exercise **A3.2**

The same results can be obtained from a probability tree diagram as shown in Figure A3.9.

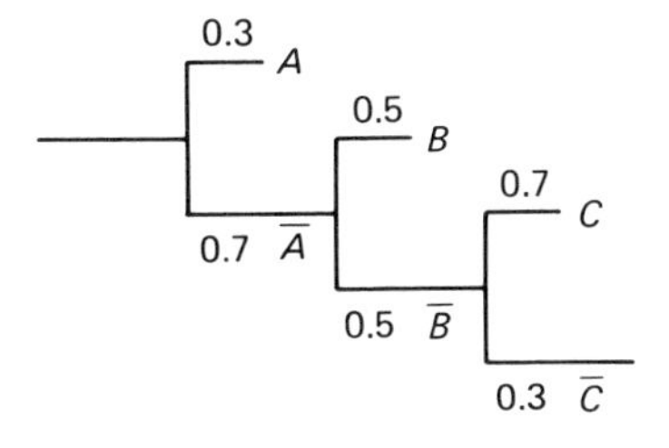

Figure A3.9 Probability tree diagram for Exercise **A3.2**

Probability that the bull has been hit

$$= 0.3 + (0.7 \times 0.5) + (0.7 \times 0.5 \times 0.7)$$
$$= 0.3 + 0.35 + 0.245$$
$$= 0.895$$

Appendix 4: Glossary of Terms

Active redundancy Two or more units work continuously; some may fail but the system stays 'up'.

Availability (instantaneous) The probability that a system will be operational at a particular instant in time.

Availability (steady-state) The steady-state value of the Availability. It can be shown to equal MUT/(MUT + MDT).

BS British Standard.

BSI British Standards Institution, 2 Park Street, London W1A 2BS.

Capability The probability that the system will perform a particular operation successfully.

CCITT International Consultative Committee for Telephony and Telegraphy.

Constant failure rate Failure rate (*q.v.*) which remains constant independently of time.

Corrective maintenance Maintenance actions intended to correct a fault.

Dependability This term usually has the same meaning as 'mission reliability' (*q.v.*).

Design life The length of time that a system is designed to survive (with interim repairs if necessary).

Expected life The length of time which a component or system can be expected to last.

Failure The IEC definition is: 'The termination of the ability of an item to perform a required function.' It is also noted that: ' "failure" is an event as distinguished from "fault" which is a state.'

Failure intensity Analogous to failure rate but applied specifically to repairable systems.

Failure rate The fraction of 'good' devices at time t which fail in a unit increment of time.

Fault *See under* Failure.

Fault coverage The percentages of total possible faults that a self-checking system can detect.

Fault tolerance The ability of a system to withstand one or more failures and still remain in a useful state.

Integrity That quality which relates to the trust which can be placed in the correctness of the information supplied by the system.

IQA Institute of Quality Assurance (UK).

Maintainability The probability that a repair action will restore a failed system to full working order within a given period of time.

MART Mean Active Repair Time.

MCT Mean Cycle Time (equals MUT + MDT).

MDT Mean Down Time.

Mean residual life The mean life of an item which has already survived a period of time t. This is in general a function of t denoted by $L(t)$. For the exponential distribution alone, $L(t)$ is a constant and equal to $1/\lambda$. $L(0)$ is by definition the MTTF.

Mission profile The set of conditions (storage, powered-up etc.) and corresponding lengths of time which a system will encounter during its life or mission.

Mission reliability The probability that a system will successfully perform a given mission.

Mission time For a system which is used continuously (such as a telephone exchange), the mission time is the design life of the system. However, for a system which is used intermittently (such as an aircraft), the mission time may be the length of time the system is in use (for example, the length of time to fly from A to B).

MRDT Mean Response Delay Time.

MTBF Mean Time Between Failures. (This term can be misleading, and it is usually better to use MTTF or MUT; however, these terms are not so generally well known.)

MTT Mean Travelling Time.

MTTF Mean Time to Failure.

MTTFF Mean Time To First Failure.

MTTR Mean Time To Repair.

MTTSF Mean Time To Second Failure.

MUT Mean Up Time.

NCSR National Centre for Systems Reliability (UK).

Passive redundancy One or more units are held in standby ready to replace the working unit.

PAT Post-repair Administrative Time.

PTT Post, Telephone & Telegraph (Authority). The regulatory authority (such as Deutsche Bundespost) which many countries have.

Redundancy The use of two or more units when only one could cope.

Reliability The probability that a device or system will operate for a given period of time and under given operating conditions.

Security The condition that results when measures are taken that protect information, personnel, systems, components and equipment from unauthorised persons, acts or influences. (Weik, M. H., *Communications Standard Dictionary*, Van Nostrand, 1983.)

System effectiveness A (sometimes crude) assessment of the 'goodness' of a system. It may be particularly useful in comparing competing systems. One definition (ARINC) is: 'The probability that the system can successfully meet an operational demand within a given time when operated under specified conditions.'

Useful life 'The period from a stated time during which under stated conditions an item has an acceptable failure intensity or until an unrepairable failure occurs' (CCITT G111).

Weibull distribution A statistical distribution which is particularly useful in analysing non-constant failure rate data.

Index